SpringerBriefs in Earth System Sciences

Series editors

Gerrit Lohmann, Bremen, Germany
Lawrence A. Mysak, Montreal, Canada
Justus Notholt, Bremen, Germany
Jorge Rabassa, Ushuaia, Argentina
Vikram Unnithan, Bremen, Germany

More information about this series at http://www.springer.com/series/10032

Manuel Enrique Pardo Echarte
Jorge Luis Cobiella Reguera

Oil and Gas Exploration in Cuba

Geological-Structural Cartography using Potential Fields and Airborne Gamma Spectrometry

 Springer

Manuel Enrique Pardo Echarte
Scientific-Research Unit Exploration
Centro de Investigaciones del Petróleo
El Cerro, La Habana
Cuba

Jorge Luis Cobiella Reguera
Universidad de Pinar del Río "Hermanos
 Saíz Montes de Oca"
Pinar del Río
Cuba

ISSN 2191-589X   ISSN 2191-5903   (electronic)
SpringerBriefs in Earth System Sciences
ISBN 978-3-319-56743-3   ISBN 978-3-319-56744-0   (eBook)
DOI 10.1007/978-3-319-56744-0

Library of Congress Control Number: 2017937296

Printed on acid-free paper

This Springer imprint is published by Springer Nature
The registered company is Springer International Publishing AG
The registered company address is: Gewerbestrasse 11, 6330 Cham, Switzerland

Foreword

For any country, it is essential to conduct an ongoing assessment of potential energy and mineral resources with a view to the renewal of its exploration policy and strategy of economic development. This is based on the increase in the acquisition of new geological, geophysical, geochemical and other data (geological knowledge) and in the constant improvement of the criteria and methods of geological exploration.

From this, the tectonic-structural regionalization with purposes of oil exploration in Cuba, focusing on the mapping of potential new targets in different regions of oil-gas interest, start from the contribution of potential fields and airborne gamma spectrometry to geological-structural cartography and geological exploration, depending both on the type of geology, climate, and topography of the investigated territory. Thus, the territory of the Republic of Cuba is privileged by its contrasting alpine geology and its tropical climate, which determines the presence of residual soils and in situ developed weathering crusts and an essentially flat relief. For others, Cuba has lifted all its territory with aeromagnetic and airborne gamma spectrometry at scale of 1:50,000 and gravimetric survey on the same scale, with 80% coverage.

Each of these methods has contributed to the study of regional and local geological constitution:

- The application of the gravimetric method offers the possibility of studying the regional geological constitution, with better results for the folded belts, such as Cuba, thus permitting the tectonic regionalization, geological-structural mapping of large units and location of structures in the sedimentary cover. This is an effective means of mapping sedimentary basins and major tectonic features with which various mineral and energy resources are linked sometimes. From a local point of view, it is accepted for locating and mapping of bodies of salt, reef, granite, and ultrabasites.
- The aeromagnetic provides an aid to geological mapping of volcanogenic-sedimentary and intrusive formations in volcanic arcs, such as Cuban territory. In the presence of nonmagnetic sedimentary rocks, aeromagnetic survey data

provide information on the nature and depth of the basic-ultrabasic and/or crystalline basement. Locally, the ability to map geological-structural features is enhanced by the ability to detect low amplitude anomalies; besides intrusive bodies (granitoid) and protrusive (ophiolites) can be distinguished, often directly.

- The airborne gamma spectrometry (AGS) offers potential for mapping and subdividing acid-medium igneous and metamorphic rocks and it highlights the rock types that are characterized by unusual amounts or very low proportions of radioelements as basic–ultrabasic complex. In less radioactive environments such as volcanogenic-sedimentary terrains and sedimentary basins, the most subtle contrasts also offer reliable guide mapping. The advantage of AGS compared to other techniques of remote sensing is in mapping soil variations in areas of dense vegetation and areas of flat land. On occurrences of oil and gas, decomposition of clays in soils product of light hydrocarbons microseepage is responsible for minima observed radiation: potassium is leached from the system toward the edges of the accumulation vertical projection where precipitates result in a "halo" of high values. The thorium remains relatively fixed in their original distribution within the insoluble heavy minerals; hence, there is an observed minimum of the ratio K/Th surrounded by maximums on these deposits. On the periphery of these anomalies local increases of U(Ra) are also observed. Finally, AGS data can complement the structural interpretation of other geophysical data, playing a major role in controlling the surface geology, where some structures that do not produce magnetic and gravity anomalous responses can be deduced from these data.

The reader will find in this Monograph an interesting and valuable overview of the geology of Cuba, a synthesis of the main results in the tectonic-structural regionalization with purposes of hydrocarbon exploration, focusing on the mapping of potential new targets in different regions of oil-gas interest. Also, this book offers information on mapping geological-structural of igneous and metamorphic rock units in different regions of Cuba; all of them from potential fields and airborne gamma spectrometry data.

Dr. Osvaldo Rodríguez Morán
Scientific-Research Unit Exploration
Department of Regional Geology
and Geological Survey
Centro de Investigaciones del Petróleo

Preface

A summary of the main results achieved in the tectonic-structural regionalization with purposes of hydrocarbon exploration, from data of potential fields and airborne gamma spectrometry is offered, focusing on the mapping of possible new oil-gas objectives in the regions of the Earth Blocks 9, 23 and 17–18. In some of the new places of interest (Majaguillar, Motembo Sureste, Guamutas and Maniabón) reconnaissance works were carried out by profiles of *Redox Complex* (complex of unconventional geophysical-geochemical exploration techniques) with positive results. Thus, Maniabón, south of the bays of Puerto Padre and Nuevitas was established as the location of main interest.

The results of the geological-structural mapping of igneous and metamorphic rocks units in the western regions (Havana-Matanzas), central (Cienfuegos-Villa Clara-Sancti Spiritus), and central-eastern (Camagüey-Tunas-Holguín) of Cuba also provide data of potential fields and airborne gamma spectrometry.

The contribution of potential fields and airborne gamma spectrometry data for geological-structural mapping of different areas of study in Cuba satisfies the well-established regularity that potential fields help, basically, to tectonic-structural deciphering of the territory and, to a lesser extent, to lithological mapping of the different units present; resulting in reverse the contribution of airborne gamma spectrometry data. Gravity data allow to identify different geological-structural features: for minimum, those associated with the Cuban North Thrust Belt, the southern metamorphic massifs, the granitic igneous bodies, the synorogenic basins and structural depressions; by highs, linked with huge thicknesses of volcanic rocks, and ophiolitic bodies; as well as by lineaments, the major tectonic boundaries within the Cuban Orogen. The aeromagnetic data allow mapping the main tectonic boundaries; the southern metamorphic massifs; the synorogenic basins and structural depressions; granitoid belts; ophiolitic bodies; and development areas of volcanic rocks. In aeromagnetics, the faculty of lithological mapping is given from the differential distribution of magnetite in various rock units. These data allow also making quantitative estimates of depth to magnetic targets. The airborne gamma spectrometry data identify, by increased values of U(Ra), units with high graphite content (organic matter), and those associated with acid igneous rocks. Increased values of potassium are linked mainly to medium-alkaline and acid igneous rocks. Thorium increased values characterize, in general, metamorphites. Some highly developed weathering mantles on ophiolites are expressed by increments of U(Ra) and Th. Hydrocarbon deposits are expressed by minimum of K/Th ratio surrounded by maximum and, at its periphery, local increases of U(Ra) is observed.

El Cerro, La Habana, Cuba   Manuel Enrique Pardo Echarte
Pinar del Río, Cuba   Jorge Luis Cobiella Reguera

Acknowledgements

We thank the institutions: Centro de Investigaciones del Petróleo e Instituto de Geología y Paleontología (Servicio Geológico de Cuba), for allowing us to publish partial information concerning various research projects.

Contents

Abbreviations

AAC	Ascendant analytic continuation
AC (CA)	Anomalous complex
ACAO	Antillean-Central America orogen
AGS	Airborne gamma spectrometry
BA-PBPg	Back-arc Basin–Piggyback Basin (Paleogene)
CB (CC)	Central Basin
CVA (AVC)	Cretaceous volcanic arc
CVR (RVC)	Cretaceous volcanic rocks
EDM	Elevation digital model
It	Integral channel in AGS
K, U (Ra) and Th	Potassium, uranium (radium) and thorium
K/Pg boundary	Cretaceous/Paleogene boundary
K–Ar	Potassium–argon
Kn	Normalized magnetic susceptibility
KVAT	Cretaceous volcanic arc terrain
Ma	Millions of years
MME	Escambray metamorphic massif
MTU	Major tectonic units
NCTB (CCNC)	Northern Cuban Thrusts Belt
NOB	Northern ophiolite belt
OB (CO)	Ophiolite belt
ORP	Redox potential
PF	Potential fields
RER	Reduced spectral reflectance
RTP	Reduced to pole
SD (DE)	Structural depression

SR-APN-E	Sierra del Rosario-Alturas de Pizarras del Norte-Esperanza unit
TSU (UTE)	Tectonic-stratigraphic units
Ur	Reduced redox potential
V, Pb and Zn	Vanadium, lead and zinc
VD	Vertical derivative

Chapter 1
The State of Art

Abstract In the application of potential fields in regional investigations, the gravimetric method solves the problems of the study of regional geological constitution, with better results for the folded belts. Regional studies that make use of gravimetric surveys also permit detailing the main tectonic features such as faults and other alignments, with which various mineral and energy resources are linked. Meanwhile, the aeromagnetic survey is helpful for geological mapping of vast regions with a thick sedimentary cover, where geological-structural features can be revealed if some magnetic horizons are present within the sedimentary sequence. In the absence of magnetic sediments, aeromagnetic survey data can provide information on the nature and form of basic–ultrabasic and/or crystalline basement. Both cases are applicable to oil and gas exploration. Regarding airborne gamma spectrometry, this provides a direct measurement of the naturally distribution of radioelements (K, U, and Th) occurring in the ground surface. Potassium is a major constituent of most rocks, while uranium and thorium are present in trace amounts, such as mobile and immobile elements, respectively. As the concentration of these radioelements varies among different types of rocks (and derived residual soils), distribution can be used to distinguish the different lithologies. Regarding hydrocarbon exploration, changes in mineral stability on columns of mineralized rocks above hydrocarbon deposits derived different types of morphometric, geophysics, and geochemistry significant responses.

Keywords Potential fields · Gravimetry · Aeromagnetics · Airborne gamma spectrometry · Geological-structural mapping · Oil and gas exploration

1.1 Introduction

In the geophysical literature, since its inception, numerous examples of the application of potential fields (gravimetry and magnetometry) for the tectonic-structural regionalization of the various territories, geological-structural mapping and geological exploration (mainly, oil and gas, and metal ores) are collected. Similarly, the

© The Author(s) 2017   1
M.E. Pardo Echarte and J.L. Cobiella Reguera, *Oil and Gas Exploration in Cuba*,
SpringerBriefs in Earth System Sciences, DOI 10.1007/978-3-319-56744-0_1

airborne gamma spectrometry applications for lithological mapping and geological exploration (minerals and hydrocarbons) are recognized. This chapter, without trying to grasp the scale of this subject, aims to provide a summary of its most importantly known aspects. Besides, the basic premises in the application of non-conventional geophysical-geochemical hydrocarbon exploration techniques are briefly addressed.

1.2 Synthesis of Potential Fields and Airborne Gamma Spectrometry

Numerous investigators (Dobrin and Savit 1988; Garland 1989; Gubins 1997; Keary et al. 2002) have described the possibilities in the application of potential fields in regional investigations. The gravimetric method solves the problems of the study of regional geological constitution, with better results for the folded belts, allowing tectonic regionalization, the geological-structural mapping of large units and search for structures in the sedimentary cover. It is an effective means of mapping sedimentary basins where basement rocks have a density consistently higher than sediment. In addition, it is suitable for localization and mapping salt, ultrabasic and reef bodies, due to the usually low and high density compared with the surrounding formations. Regional studies that make use of gravimetric surveys also permit detailing the main tectonic features such as faults and other alignments, with which various mineral and energy resources are linked. Meanwhile, the aeromagnetic survey is helpful for geological mapping of vast regions with a thick sedimentary cover, where geological-structural features can be revealed if some magnetic horizons, such as ferruginous sandstones and shales or tuffs and lava flows, are present within the sedimentary sequence. In the absence of magnetic sediments, aeromagnetic survey data can provide information on the nature and form of basic–ultrabasic and/or crystalline basement. Both cases are applicable to oil and gas exploration, for the location of structural traps within the sediments or linked to features of the topography of the basement that could influence the overlying sedimentary sequence. In aeromagnetics, capacity of mapping geological-structural features is enhanced by the possibility of mapping low amplitude anomalies. Intrusive (granitoids) and protrusive (ophiolites) bodies often can be distinguished directly based on magnetic observations.

According to Verduzco et al. (2004), the tilt derivative of the total magnetic field and its total horizontal derivative are useful for structural mapping and mineral exploration. The tilt derivative of the field is defined as

$$TDR = ArcTan\left(\frac{VDR}{THDR}\right),$$

where VDR and THDR are the first vertical derivative and the total horizontal derivative, respectively, of the total magnetic intensity T (reduced to pole)

$$VDR = \frac{dT}{dz}$$

$$THDR = \sqrt{\left(\frac{dT}{dx}\right)^2 + \left(\frac{dT}{dy}\right)^2}$$

The total horizontal derivative of the tilt derivative is defined as

$$HD\_TDR = \sqrt{\left(\frac{dTDR}{dx}\right)^2 + \left(\frac{dTDR}{dy}\right)^2}$$

The chains axis of maximums of this attribute are coincident with structural limits or tectonic alignments. Subsequently, other researchers (Fairhead et al. 2009; Ghosh and Dasgupta 2013; Arisoy and Dikmen 2013) have offered in their work, further details on these derivative parameters.

Regarding airborne gamma spectrometry (AGS), according to Darnley and Ford (1989), Graham and Bonham-Carter (1993), Grasty and Shives (1997), this provides a direct measurement of the natural distribution of radioelements (K, U, and Th) occurring in the ground surface. Potassium is a major constituent of most rocks, while uranium and thorium are present in trace amounts, such as mobile and immobile elements, respectively. As the concentration of these radioelements varies among different types of rocks (and derived residual soils), distribution can be used to distinguish the different lithologies. The method shows potential for mapping and subdivides acid igneous and metamorphic rocks and highlights the rock types that are characterized by unusual amounts or very low proportions of radioelements such as basic–ultrabasic complex. Applications have been expanded in the less radioactive environments such as sedimentary basins, volcanogenic-sedimentary terrains and weathered areas, which in its subtle contrast guides offer reliable mapping.

According to Price (1985), about the occurrences of oil and gas, decomposition of clay in the soil product of light hydrocarbons microseepage it is responsible for the minimum observed radiation: potassium is leached from the system to the edges of the vertical projection of accumulation, which precipitates resulting in a "halo" of high values. The thorium remains relatively fixed in their original distribution within the insoluble heavy minerals; hence, there are observed minimum of the K/Th ratio surrounded by a maximum on these deposits. On the periphery of these anomalies are also observed local increases of U(Ra).

The mapping of lithological units (and/or derived residual soils) is affected by many factors: the contrasts of radioelements content among lithological units; the degree of exposure of bedrock and soil cover; the measurement geometry (relief); the relative distribution of transported and in situ soils; the nature and type of weathering; the moisture content of the soil; and the vegetation cover. However, the AGS shows advantages compared to other techniques of remote sensing in mapping soil variations in areas of dense vegetation and areas of flat land.

From the point of view of interpretation, composite images provide simultaneous display of up to three parameters in an image and facilitates correlation and zonation based on subtle differences in the numerical values. The following combinations are developed by the US Geological Survey (USGS):

1. The radioelements composite image that combines data of K (red), Th (green), and U (blue).
2. The composite image that combines data of K (red), with ratios K/Th (green) and K/U (blue).
3. The composite image that combines data of U (in red), with ratios U/Th (green) and U/K (blue).
4. The composite image that combines data of Th (red), with ratios Th/U (green), and Th/K (blue).
5. The composite image of radioelements relations that combines data from the three radiometric relations U/Th (red), U/K (green), and Th/K (blue).

The composite image of radioelements offers many advantages in terms of lithological discrimination based on color differences in them and highlights areas where, in particular, each radioelement has an absolute or relative higher concentration.

Furthermore, AGS data can complement the structural interpretation of other geophysical data, playing a major role in controlling the surface geology where some structures that do not produce magnetic and gravity anomalous responses can be deduced from these data.

1.3   Synthesis of Geological-Geophysical-Geochemical Aspects

According to Price (1985) and, Pardo Echarte and Rodríguez Morán (2016), geological assumptions underlying the application of unconventional geophysical-geochemical hydrocarbon exploration techniques (such as **Redox Complex**) and that, in turn, argue the design of geophysical-geochemical-morphometric interpretive scenarios, that are as follows:

- The "Reducing Chimneys" are columns of mineralized rocks above hydrocarbon deposits which were modified by the vertical migration of light hydrocarbons and/or any other association of reduced species (metal ions) which "oxidize" from microbial action, to create a reducing environment.
- The main products of the microbial hydrocarbon oxidation (CO_2) and microbial reduction of sulfur (H_2S), drastically change the pH/eH of the system. Changes in pH/eH result in changes of the mineral stability:

– Precipitation of various carbonates.
– Decomposition of clays (therefore, the concentrations of silica and alumina increase).
– Precipitation of magnetite/maghemite, iron sulfides (such as pyrrhotite and griegita) or coprecipitation of iron and/or manganese with calcite in carbonate cements on hydrocarbon deposits.

Morphometric, geophysics, and geochemistry response to changes in the previous mineral stability is as follows:

- Secondary mineralization of calcium carbonate and silicification, resulting in denser and erosion resistant surface materials (formation of positive geomorphic anomalies and resistivity maximums).
- The decomposition of clay is responsible for the radiation lows reported on oilfields; potassium is leached from the system (causing a central low) to the edges of the hydrocarbon deposit vertical projection, which precipitates resulting in a "halo" of high values. The thorium remains relatively fixed in their original distribution within the insoluble heavy minerals; hence, a minimum of K/Th ratio surrounded by peaks over oil–gas deposits is observed. On the periphery of these anomalies, local increases of U(Ra) are observed.
- The conversion of nonmagnetic iron ores (oxides and sulfides) in more stable magnetic varieties results in an increase of magnetic susceptibility correlatable with minimal of redox potential (ORP) which justifies integrating both techniques within the **Redox Complex**. Induced polarization anomalies are also observed.
- The arrival to the surface of the metal ions contained in hydrocarbons (V, Ni, Fe, Pb, and Zn, among others) determines the presence of a subtle anomaly of these elements in the soil and thereof a slight change in color. The latter is reflected by spectral reflectance anomalies, fact justifying the integration of these techniques within the **Redox Complex**.

1.4 Conclusions

In the application of potential fields in regional investigations, the gravimetric method solves the problems of the study of regional geological constitution, with better results for the folded belts. Regional studies that make use of gravimetric surveys also permit detailing the main tectonic features such as faults and other alignments, with which various mineral and energy resources are linked. Meanwhile, the aeromagnetic survey is helpful for geological mapping of vast regions with a thick sedimentary cover, where geological-structural features can be revealed if some magnetic horizons are present within the sedimentary sequence. In the absence of magnetic sediments, aeromagnetic survey data can provide information on the nature and the form of basic–ultrabasic and/or crystalline basement. Both cases are applicable to oil and gas exploration.

Regarding airborne gamma spectrometry, this provides a direct measurement of the natural distribution of radioelements (K, U, and Th) occurring in the ground surface. Potassium is a major constituent of most rocks, while uranium and thorium are present in trace amounts, such as mobile and immobile elements, respectively. As the concentration of these radioelements varies among different types of rocks (and derived residual soils), distribution can be used to distinguish the different lithologies. The method shows potential for mapping and subdivides acid igneous and metamorphic rocks and highlights the rock types that are characterized by unusual amounts or very low proportions of radioelements such as basic–ultrabasic complex.

Regarding hydrocarbon exploration, changes in mineral stability on columns of mineralized rocks above hydrocarbon deposits derived different types of morphometric, geophysics, and geochemistry significant responses:

- Secondary mineralization of calcium carbonate and silicification, resulting in denser and erosion resistant surface materials (formation of positive geomorphic anomalies and resistivity maximums).
- The decomposition of clay is responsible for the radiation lows reported on oilfields: potassium is leached from the system to the edges of the hydrocarbon deposit vertical projection, which precipitates resulting in a "halo" of high values. The thorium remains relatively fixed in their original distribution within the insoluble heavy minerals; hence, a minimum of K/Th ratio surrounded by peaks over oil–gas deposits is observed. On the periphery of these anomalies, local increases U(Ra) are observed.
- The conversion of nonmagnetic iron ores (oxides and sulfides) in more stable magnetic varieties results in an increase of magnetic susceptibility correlatable with minimal of redox potential (ORP). Induced polarization anomalies are also observed.
- The arrival to the surface of the metal ions contained in hydrocarbons (V, Ni, Fe, Pb, and Zn, among others) determines the presence of a subtle anomaly of these elements in the soil and thereof a slight change in color, which it is reflected by spectral reflectance anomalies.

References

Arisoy MÖ, Dikmen Ü (2013) Edge detection of magnetic sources using enhanced total horizontal derivative of the tilt angle. Bulletin of the earth sciences application and research centre of Hacettepe University. Yerbilimleri 34(1):73–82
Darnley AG, Ford KL (1989) Regional airborne gamma-ray survey: a review. In: Proceedings of exploration 87: third decennial international conference on geophysical and geochemical exploration for minerals and ground water, Geol Surv Canada, Special vol 3, p 960

Dobrin MB, Savit OH (1988) Introduction to geophysical prospecting. McGraw Hill international editions, 4th edn, pp 867

Fairhead JD, Salem A, Williams SE (2009) Tilt-depth: a simple depth-estimation method using first order magnetic derivatives. Search and discovery article #40390 (Adapted from poster presentation at AAPG International Conference and Exhibition, Cape Town, South Africa, October 26–29, 2008)

Garland GD (1989) Proceedings of exploration 87. Third Decennial international conference on geophysical and geochemical exploration for minerals and groundwater, Special vol 3, Ontario Geological Survey, pp 914

Ghosh GK and Dasgupta R (2013) Edge detection and depth estimation using 3D Euler deconvolution, Tilt angle derivative and TDX derivative using magnetic data of thrust fold belt area of Mizoram. 10th Biennial International Conference & Exposition, p 070

Graham DF, Bonham-Carter GF (1993) Airborne radiometric data: a tool for reconnaissance geological mapping using a GIS. Photogram Eng Remote Sens 58:1243–1249

Grasty RL, Shives RBK (1997) Applications of gamma ray spectrometry to mineral exploration and geological mapping, Workshop presented at exploration 97. Fourth Decennial Conference on Mineral Exploration

Gubins AG (1997) Proceedings of exploration 97. Fourth Decennial International Conference on Mineral Exploration. Prospectors and Developers Association of Canada, pp 1065

Keary P, Brooks M, Hill I (2002) An introduction to geophysical exploration, 3rd edn. Blackwell Science Ltd., England 262 pp

Pardo Echarte ME, Rodríguez Morán O (2016) Unconventional methods for oil & gas exploration in Cuba. Springer Briefs in Earth System Sciences. doi:10.1007/978-3-319-28017-2

Price LC (1985) A critical overview of and proposed working model for hydrocarbon microseepage. U.S. Department of the Interior Geological Survey. Open-File Report, pp 85–271

Verduzco B, Fairhead JD, Green CM, McKenzie C (2004) New insights into magnetic derivatives for structural mapping. Lead Edge 23(2):116–119

Chapter 2
Overview of the Geology of Cuba

Abstract The Cuban Orogen can be divided into two major structural and stratigraphic units: basement and cover. The basement is the mega complex of igneous, metamorphic, and sedimentary rocks that lies below the little deformed section of the cover. It is divided into several tectonic large units, according to its structural style and age of the rocks. We can distinguish three large complexes: (a) Proterozoic basement, (b) Mesozoic basement, and (c) Paleogene folded belt. The Proterozoic basement outcrops in very limited areas and its structure is unclear. The Mesozoic basement consists of four complexes of very different nature: the Mesozoic paleomargin of the SE North American plate, containing Jurassic-Cretaceous sequences with varying degrees of deformation; the remaining three units, the ophiolite association, successions of volcanic arcs, and southern metamorphic terrains, have traits of tectono stratigraphic terrains. The links between the four major structures of the Paleogene deformed belt are much clearer. However the outstanding deformations and horizontal transport suffered by some units, the primary spatial relationships (paleogeographic) between them are essentially preserved. At the Paleogene folded and faulted belt are distinguished: • Foreland basin successions. • Piggyback basins successions. • Sierra Maestra-Cresta Caimán volcanic arc. • Synorogenic basin of Middle and Upper Eocene S Eastern Cuba. The cover composed of deposits from Lower or Middle Eocene to Quaternary, comprises the younger deposits of the section, little deformed in relation to the underlying layers, usually separated from these by remarkable structural discordance. Throughout Cuba, the cover is completely devoid of evidences of magmatic, metamorphic, and hydrothermal activity.

Keywords Cuban orogen · Base · Cover · Mesozoic continental paleomargin · Mesozoic ophiolite association · Cretaceous volcanic arcs terrain · Paleogene piggyback basins · Cuban foreland basin

© The Author(s) 2017   9
M.E. Pardo Echarte and J.L. Cobiella Reguera, *Oil and Gas Exploration in Cuba*,
SpringerBriefs in Earth System Sciences, DOI 10.1007/978-3-319-56744-0_2

2.1   Introduction

Cuba and its surroundings are a geological mosaic belonging to the southeastern portion of the North American plate. In the mosaic are rocks whose age ranges from Proterozoic to Quaternary, formed in various geological contexts. Much of Precenozoic registration is form by rocks originated at a considerable distance from their current location. Cuban territory is part of the Antillean-Central America orogen (ACAO), which extends from northern Central America to the Virgin Islands, located east of Puerto Rico. From the Middle Eocene, the deformed belt, located immediately at the border of North American-Caribbean plates, has been dissected by numerous faults that have generated large marine depressions as the Cayman trough and the Yucatan basin, which significantly impede the decipherment of their structure and history. Today the northern part of ACAO, where Cuba lies, belongs to the North American plate, while the south (where lie the other Greater Antilles) is part of the Caribbean plate.

   This chapter aims to provide an overview of the geology of Cuba, particularly of the central western, central, and center eastern regions of the territory, with which the research papers presented in Chaps. 3 and 4 relate to. Its content aims to satisfy the geological interpretation of the various issues raised in the geophysical interpretation of potential and airborne gamma spectrometry fields in the mentioned chapters.

2.2   The Cuban Orogenic Belt

The ACAO in Cuba can be divided into two major structural and stratigraphic units described here as basement and cover (fold belt and Neoauthocton, sensu Iturralde-Vinent 1996a, 1997). The basement (Cobiella-Reguera 2000) is defined as the mega complex of igneous, metamorphic, and sedimentary rocks that lies below the little deformed section of the cover. The cover consists of deposits of Lower or Middle Eocene to Quaternary (west of Yabre lineament-Y); Middle Eocene to Quaternary (between this lineament and the Guacanayabo-Nipe lineament-GN) and Upper Eocene high to Quaternary in Eastern Cuba, the south and east of GN lineament. Throughout Cuba, the cover is completely devoid of evidence of magmatic, metamorphic, and hydrothermal activity.

   Cuban base is divided into several tectonic larger units, according to its structural style and age of the rocks. We can distinguish three large complexes: (a) Proterozoic basement, (b) Mesozoic basement (c) Paleogene folded belt.

   The Proterozoic basement outcrops are very limited and its tectonic style is unclear.

   The Mesozoic basement consists of four complexes of very different nature: the Mesozoic paleomargin of the SE North American plate presents Jurassic-Cretaceous sequences with varying degrees of deformation; the remaining three, the ophiolite association, Cretaceous volcanic arcs, and the southern metamorphic terrains, have traits of tectonostratigraphic terrains. Southern metamorphic massifs may be considered proximal terrains, as their cuts show a clear stratigraphic link with the Mesozoic paleomargin of southeastern North America.

The links between the four regional structures of the Paleogene deformed belt are much clearer and, notwithstanding the considerable deformations and horizontal transport suffered by some, the primary spatial relationships (paleogeographic) between them are essentially preserved. At the Paleogene folded and faulted belt are distinguished:

- Foreland basin successions.
- Piggyback basins successions.
- Sierra Maestra-Cresta Caimán volcanic arc.
- Synorogenic basin of Middle and Upper Eocene S Eastern Cuba.

The Eocene-Quaternary cover comprises younger deposits of the stratigraphic section, little disjointed in relation to the underlying layers, usually separated from these by remarkable unconformities.

Basement

The basement meets an extremely varied set of rocks of different composition and ages, located below the Eocene-Quaternary sedimentary cover.

2.2.1 Proterozoic Basement and Jurassic Granites

The Precambrian basement is of great interest to the Antillean regional geology. In Cuba, it is known from small outcrops of Precambrian phlogopitic marbles with Proterozoic radiometric ages (903 and 952 million years ago) to the east and north of Motembo, Villa Clara province. These radiometric ages fall within the geochronological interval corresponding to the greenvillian orogeny, recorded in the southeast North American plate. At Socorro (Matanzas), these marbles are possibly cut by Jurassic granites with an age (U-Pb) of 172 million years (Renne et al. 1989; 139 and 150 Ma, according to previous dating of Somin and Millán 1981) and covered by an arkosic paleosoil (Pszczółkowski 1986). The paleosoil lies beneath the Upper Jurassic Constancia Formation, the marbles are the only known Precambrian rocks in the Greater Antilles. The discovery of marble and granite clasts in the nearby Jurassic arkose (Pszczółkowski and Myczyński 2003) is a strong indicator of a basement with Precambrian rocks for the southern edge of the North American plate.

Moreover, in areas near the North American Plate (SE Gulf of Mexico, Belize-Yucatan and Florida–Bahamas platform) Paleozoic rocks are known. The presence of Paleozoic fossils redeposited in Jurassic rocks of the Guaniguanico mountain range also suggests the presence of layers of that age in the Premesozoic basement.

2.2.2 Mesozoic Basement

The Mesozoic basement is the most complicated basement element formed by four large complexes of varied composition, age, and origin, separated by tectonic boundaries. These complexes are:

- North American paleomargin.
- Mesozoic ophiolite association.
- Cretaceous volcanic arcs terrain (KVAT), including its metamorphic basement and Campanian-Maastrichtian sedimentary cover).
- Southern metamorphic terrains.

2.2.2.1 North American Mesozoic Paleomargin

They form a diverse group of Jurassic and Cretaceous sedimentary sequences accumulated in an extensional continental margin, with some mafic tholeiitic magmatic bodies.

We can consider four different sections, related to different structures of the Mesozoic margin of the North American plate. From west to east they are:

a. Cordillera de Guaniguanico, linked to the Yucatan platform and the southeast Gulf of Mexico.
b. Sections between Havana and Camagüey, linked to the Florida–Bahamas platform.
c. Sections from northwest Holguin (Gibara), linked to the southeast of the Florida–Bahamas platform.
d. Sections from easternmost Maisí, possibly related to the Florida–Bahamas platform (Southeastern border), affected by an episode of regional metamorphism.

In each of them, the North American Mesozoic paleomargin is divided into smaller units, characterized by their own structural position, stratigraphy, and tectonic style (Meyerhoff and Hatten 1968; Pardo 1975; Pszczółkowski 1978; Iturralde-Vinent 1996a; Cobiella-Reguera 2000; Pszczółkowski and Myczyński 2003; Linares Cala et al. 2011, among others).

For the purposes of this study, the most interesting cuts of North American Mesozoic paleomargin are concentrated between Havana and Holguin. Therefore, only they will be treated in this text.

Havana-Holguin sections, linked to the Florida–Bahamas platform

From Havana, the paleomargin stratigraphy presents significant changes compared to Western Cuba. This is the region (Bahamas paleomargin, sensu Iturralde-Vinent 1996a) where the concept of zones or structure-facial units, widely used in Cuban geology (Khudoley and Meyerhoff 1971; Pszczółkowski 1982; Echevarría-Rodríguez et al. 1991; Linares Cala et al. 2011), present the broadest and robust development. In outcrops and wells, from north to south, one can distinguish the following tectonostratigraphic units of North American paleomargin:

- Cayo Coco.
- Remedios (R).
- Camajuaní (C).
- Placetas (P).

The essential features of each (sometimes with different names) are discussed in various publications (Meyerhoff and Hatten 1968; Khudoley and Meyerhoff 1971; Pardo 1975; Iturralde-Vinent 1996a; Cobiella-Reguera 2009; Linares Cala et al. 2011, among others).

In the territory between the meridian 83° 30′ W and Yabre lineament (in the provinces of Havana, Mayabeque, and Matanzas), outcrops of the paleomargin rocks are limited. However, data from wells near the north coast and some brachianticlines in northern Matanzas indicate that, under the nappes of the Cretaceous volcanic successions and the rocks of the ophiolite association that form the core of the so-called "Havana-Matanzas anticline", must lie the Jurassic- Cretaceous paleomargin rocks. These mainly belong to Placetas unit, to a lesser extent, the Camajuaní units (latter only found in depth in this area) and Cayo Coco units (evaporites, Fig. 2.1). Further east from Yabre lineament, paleomargin outcrops are arranged rather continuous throughout North Cuba to circa 76° W (Holguin).

Remedios unit usually appears in the northernmost cuts. It is integrated by sections of carbonate rocks with thick to massive layering, accumulated mostly in shallow water, representing the thick banks of the southern edge of the Florida–Bahamas platform. The Remedios unit outcrops in Central Cuba, east to Yabre lineament, but is known for deep wells in the Hicacos peninsula (Matanzas), south of

Fig. 2.1 Layers of gypsum of San Adrian Formation, of possible Jurassic age, in the homonymous quarry in NW Matanzas. At this point, tectonic inclusions of ophiolitic association rocks, Cretaceous volcanic terrains and North American paleomargin are presented

Fig. 2.2 Northern paleomargin successions. The photo on the *left* corresponds to deposits with turbidite traits of the Upper Jurassic Camajuaní unit, near Zulueta, Villa Clara. On the *right*, massive rocks of Cretaceous banks Remedios unit, south of that city

the Straits of Florida. East of Chambas (Ciego de Ávila), Remedios cut contains an intermediate stratified package of carbonated turbidites (Vilató Fm., Albian–Cenomanian), which divides it into three parts.

Further south of Remedios platform and, tectonically superimposed, outcrop tectonic scales of well-stratified successions of the Camajuaní unit (Fig. 2.2) which, in essence, represent the slope layers of the Remedios platform edge. The Camajuaní unit is not known (at least on the surface) to east of La Trocha lineament. The southern portion of the North American paleomargin in Central Cuba is represented by well-stratified deep basinal sections of Placetas unit (Profile 7–8, Fig. 2.3), dated between Upper Jurassic and the K/Pg boundary. They tectonically rest on the Camajuaní unit and their stratigraphy is essentially similar to the Sierra del Rosario-Alturas de Pizarras del Norte-Esperanza unit (SR-APN-E) nappes of the Guaniguanico mountain range (Western Cuba). However, in Placetas unit, siliciclastic deposits, underlying the clay carbonated-carbonated succession of Upper Jurassic in northern Villa Clara, are under thin arkosic cuts (Pszczółkowski and Myczyński 2003), while those in Western Cuba are much more siliceous, formed in deltaic conditions (San Cayetano Formation and equivalent). This implies a pre-Tithonian tectonic differentiation between the two regions. The Placetas unit (something metamorphosed) outcrops in the "Esmeralda Complex" and inside of Veloz, Santa Teresa and Carmita formations (Upper Jurassic to Cenomanian) northeast of the city of Camagüey. In this unit there are scarce Upper Jurassic mafic magmatism. In the Sierra de Camaján (Camagüey), the base of the tectonostratigraphic unit Placetas is represented by pillow basalts, interspersed with thin layers of hyaloclastites (radiometric age 146 ± 6 Ma, Tithonian) and, to a lesser extent, of tuffites and fossiliferous tuffitic limestone.

East of La Trocha lineament, outcrops of North American northern paleomargin are mostly limited to the thick layers, usually massive, of Remedios unit. Sporadically, further north, are presented isolated outcrops of Cayo Coco unit, including possible evaporites Jurassic age of the Punta Alegre Formation (Meyerhoff and Hatten 1968;

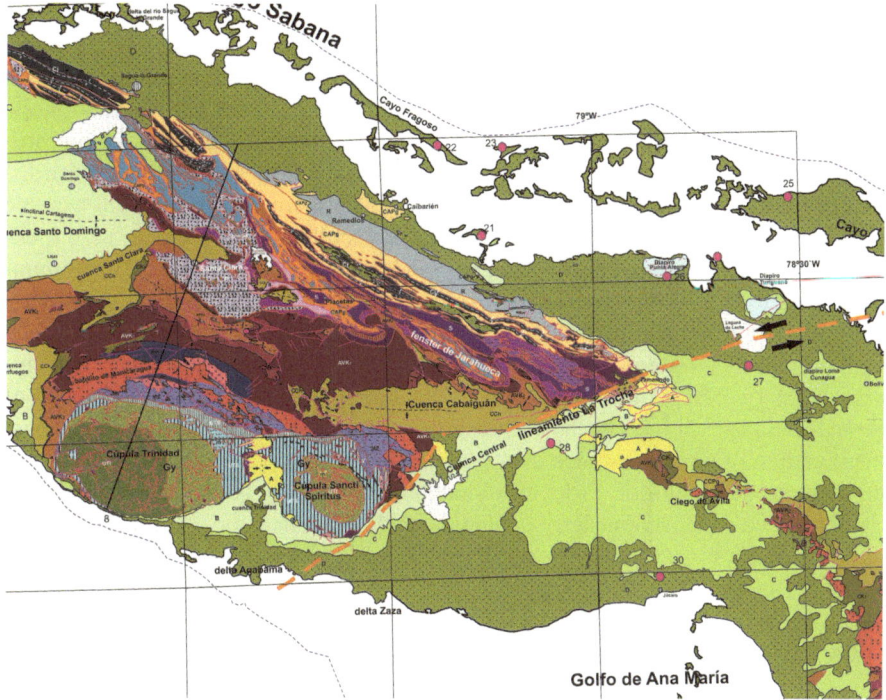

Fig. 2.3 Profile 7–8 on the Tectonic Map of Cuba (Cobiella-Reguera 2016)

Iturralde-Vinent and Roque Marrero 1982). In northwest Holguin province appear, immediately west of Gibara, the continuation of Camagüey sections, massive carbonate rocks of Remedios unit (Gibara Formation).

2.2.2.2   Mesozoic Ophiolite Association (Northern Ophiolite Belt-NOB)

The ophiolite rocks are formed by the oceanic lithosphere tectonically emplaced on the continental margins or island arcs. In the Cuban case, the ophiolitic rocks have ages between the Upper Jurassic and Early Cretaceous (Fonseca et al. 1990; Cobiella-Reguera 2005). Serpentinized ultramafites, serpentinite, mafic-ultramafic cumulative complexes and mafic rocks (volcanic and intrusive) represent the rocks of the ophiolite association. In the legend of the tectonic map of Cuba (Cobiella-Reguera 2016), the following members are distinguished: 1—volcanic-sedimentary successions, 2—shallow gabroides, 3—mafic and ultramafic cumulates, 4—moderate pressure amphibolites, 5—high-pressure amphibolites, 6—serpentinized ultramafites and serpentines. From 1 to 5, are remains of an ancient oceanic crust, while 6 are remains of the upper mantle.

The northern ophiolitic belt is well represented in Central Cuba, where its relationships with other units are more evident. The ophiolites generally cover the rocks of Placetas unit outcropping in tectonic windows (e.g., Jarahueca) and semi windows. The belt can locally overlie layers of the Camajuaní unit. This suggests that, in its movement toward the north northeast during the Early Paleogene orogeny, the ophiolitic rocks of Central Cuba could reach (at least in some territories) the old bank of Remedios area. Further east, in northern Camagüey, the tectonic map clearly shows how the Remedios unit was overwhelmed by the ophiolitic rocks on its location to the north. In Camagüey, it is also remarkable the relative abundance of gaboides beds and banded or laminated ultramafites, accompanied by chromitites deposits (profiles 9–10 and 11–12, Fig. 2.4).

In Maniabón (western Holguin province), the ophiolites are extremely dismembered (Kozary 1968; Andó et al. 1996). This, coupled with a hilly relief in most of the area and the extreme scarcity of subsurface data, further complicates the tectonic interpretation of such a complex area (Kozary 1968; Knipper and Cabrera 1974). In our opinion, the ophiolitic rocks in Maniabón form a scaly complex closely imbricated to a volcanic-sedimentary mélange, with moderate dip to the south. The ophiolites/KAVT relationship is best defined in the mountains of northeast eastern

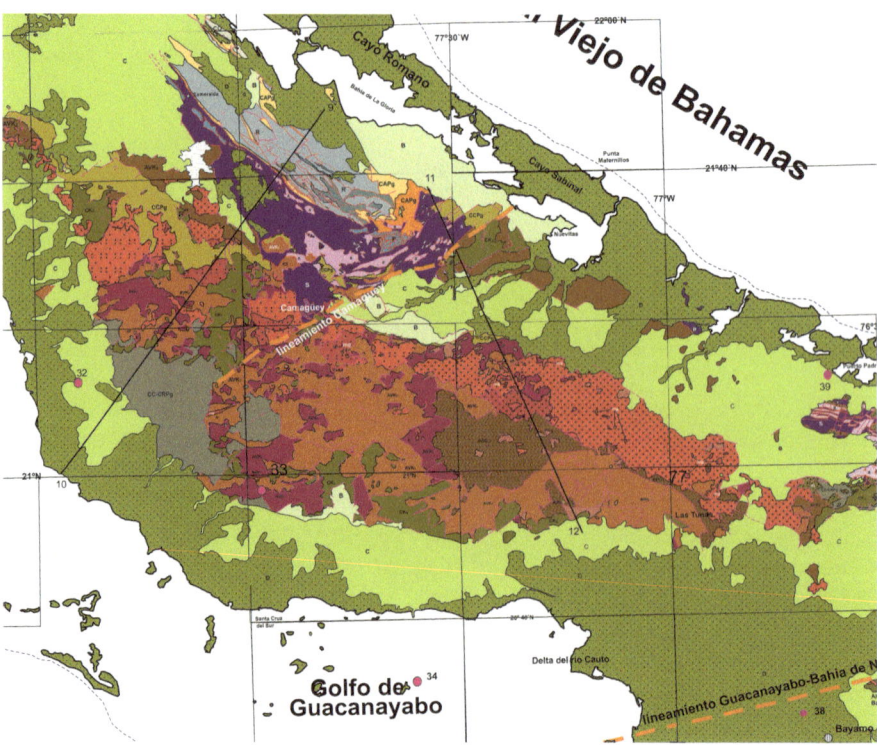

Fig. 2.4 Profiles 9–10 and 11–12 in the Tectonic Map of Cuba (Cobiella-Reguera 2016)

Cuba, where the ophiolite slab emplaced at the Maastrichtian end remains much more integrated. Such inversion of KAVT/ophiolite belt spatial relationships with respect to Western and Central Cuba distinguishes both areas (Cobiella-Reguera et al. 1984; Iturralde-Vinent 1996b; Cobiella-Reguera 2005) from the ophiolitic rocks emplaced more to the west.

There is not much quantitative information on the magnitude of displacement of the northern ophiolite belt. The 50–70 km to the north northeast displacement of the Placetas unit in Central Cuba by Pszczółkowski (1983) serves as a minimum estimate for the translation of the NOB. For the Camagüey massif was calculated 20 km (minimum) (profiles 9–10, 11–12, Fig. 2.4). For ophiolites west of Havana, a minimal displacement of 26–32 km (Cobiella-Reguera 2009) was calculated. In eastern Cuba, with a less complex geological situation, it can be estimated about 60 km of northward displacement (minimum) for the rocks of Sagua-Baracoa massif. Cobiella-Reguera (2009) considered a minimum translation to the N of 22.5 km for the ophiolites of Maniabón (profile 13–14 Fig. 2.5). In that region, it was described an elongated block of an ortogneiss of granodiorite origin, embedded in serpentinite with a K-Ar age of 196 million years (Jurassic-Sinemurian), which may reflect the minimum age of the protolith. It is considered that this block could be a basement scale of North American paleomargin, ripped off by the ophiolitic massif. All this indicates that the northern ophiolitic belt is considerably shifted to the N of its "roots" and it is not an "ophiolite suture" that marks an ancient subduction zone, as some authors have supposed.

2.2.2.3 Cretaceous Volcanic Arc Terrain (KVAT)

In much of Cuba, structurally located on ophiolitic rocks and occupying in surface a more southerly position, is arranged the terrain of Cretaceous volcanic arcs (*KVAT*), formed by Cretaceous volcanic and volcanic-sedimentary rocks, its metamorphic substrate and an Upper high Cretaceous sedimentary cover. The tectonic nature of their basal contact and the absence, until Campanian, of stratigraphic relationships with pre-Maastrichtian rocks of the remaining Mesozoic tectonic units, allow considering this set of volcanic-sedimentary rocks of Cuba as a tectonostratigraphic terrain (Blein et al. 2003). It forms part of the great Cretaceous magmatic complex that extends throughout the Greater Antilles to the Virgin Islands and possibly covers southern Guatemala. The most complete and better-outcropped area of the Cretaceous volcanic arcs is presented in the central provinces (Villa Clara, Cienfuegos, and Sancti Spiritus). In this region, the lower structural position is occupied by the Mabujina Complex amphibolite rocks. It is a diverse group of metamafics and some ultramafics are cut by granitoid intrusives, which tectonically contact with the metamorphic Guamuhaya (Escambray) massif. There is some consensus to consider this contact as a collisional suture continental margin (Escambray metamorphic massif, high-pressure/low-temperature metamorphism)—volcanic island arc (Mabujina Complex, low/medium-pressure high-temperature metamorphism). On the other hand, some experts suggest that this might be a

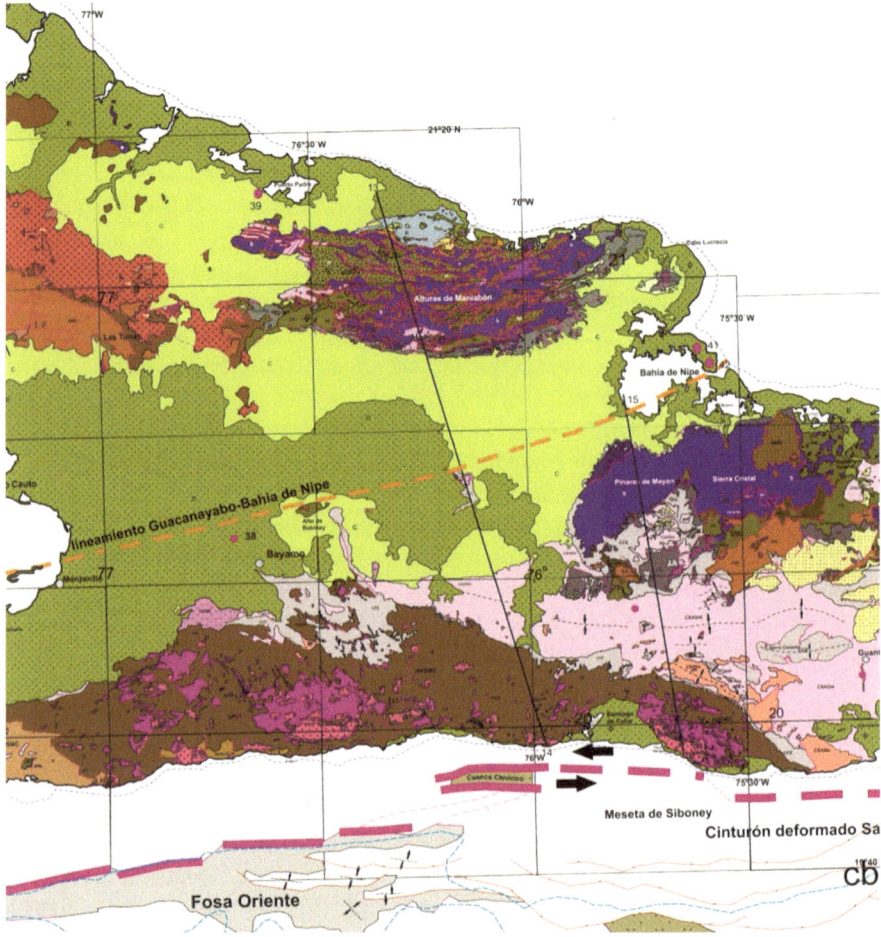

Fig. 2.5 Profile 13–14 in the Tectonic Map of Cuba (Cobiella-Reguera 2016)

distensional fault, due to the rise of Guamuhaya massif and gravitational "delam-ination" of overlying rocks (meaning, Mabujina Complex) that slid on its flanks (Pindell et al. 2006) during the Paleocene.

The Mabujina Complex is cut by numerous intrusions of various sizes. The oldest, with radiometric ages (U-Pb zircon) between 133 (Valanginian) and 110 Ma (Albian), are plagiogranites with gneissic structure and have been metamorphosed together with the amphibolites in a collisional event occurred between 90 and 88 Ma (Turonian–Coniacian). A younger generation, dated between 87 and 80 Ma (Coniacian–Campanian), corresponds to the so-called Manicaragua batholith (Stanek et al. 2009), which also injected successions of KVAT. According Stanek et al. (2006), in both Mabujina amphibolites and the Yayabo amphibolites, small pegmatite intrusive bodies are observed as well as with quartz and mica veins and milky quartz. These bodies have no ductile deformation and are post-metamorphic. For the pegmatites is recorded ages between 88 and 80 Ma (Coniacian to Middle Campanian). Therefore, the metamorphic event is prior to such record.

Above Mabujina Complex lie the bimodal volcanics Los Pasos and Porvenir formations, the last with metamorphism of greenschist facies and possible prealbian Cretaceous age. Its boundaries are tectonic with amphibolites and are cut by intrusives also injected into the Mabujina Complex. On this complex, there are the volcanic-sedimentary cuts of two volcanic island arcs (Aptian?-Albian, the lower, and Cenomanian-Campanian high, the youngest), separated by a package rich in sedimentary rocks (Provincial Fm, Albian-Cenomanian, and other more local units, Kantchev et al. 1978; Fig. 2.3). In the Tectonic map of Central Cuba (Cobiella-Reguera 2016) this situation is very clear, but it is not so clearly observed on the west and east of the region. Some differences in the chemical and petrographic composition of both arcs have been found (Draper and Barros 1994; Stanek et al. 2009), a fact that has been tried to explain in different ways (Cobiella-Reguera 2000; Proenza et al. 2006). Increasing of siliciclastic component in the Upper Cretaceous arc and the presence of some calc-alkaline rocks enriched in potassium are also remarkable. However, in general, no significant structural discordance between the two arcs is recorded.

Eastward, in the Ciego de Avila–Camagüey–Las Tunas territory, the KVAT has abundant potassic volcanics of alkaline trend in Camujiro (considered Albian-Cenomanian or Turonian) and, to a lesser degree, Piragua formations (considered Coniacian-Campanian). The overlap in their distribution areas and the evidence of a coeval volcanism with the accumulation of Piragua allow considering the possibility that this is about (at least partially) different facies of the of Late Cretaceous volcanic arc deposits (Fig. 2.4). Thus, Camujiro corresponds to the volcanic apparatus facies and its vicinity, while Piragua belongs to volcanic-sedimentary facies peripheral to volcanic foci, with terrigenous contributions from volcanic arc erosion. Further south, away from volcanoes, well-stratified beds of Aguilar Fm., considered Santonian, were deposited. North of the central belt of granitoids volcanics are calc-alkaline (Caobilla Fm., Coniacian?-Santonian?) and largely consist of pyroclastic. A fifth type, also with some presence of rocks with alkaline trend occurs in the Eastern part of Camagüey and Las Tunas, consisting of the Guaimaro Fm. (Aptian? -Cenomanian?) of basaltic rocks, and Martí Fm. (Campanian) formed by lavas and pyroclastic rocks, with some interbedded sedimentary, partially originated in subaerial conditions.

In Camagüey four groups of granitoids are distinguished:

- Granodiorite complex
- Granosienite complex
- Plagiogranite complex
- Leucocratic alkaline granites complex (Maraguán granites).

Additionally, stand out subvolcanic intrusive bodies, many possible volcanic necks. They are particularly abundant within Camujiro and Piragua formations, but also intrude the granitoids (Fig. 2.4).

There is little definition of the age of the volcanic-sedimentary successions between Ciego de Ávila and Las Tunas. Unlike Central Cuba, the absence of beds with Lower Cretaceous proven age is remarkable. Only in a oil well near the city of Camagüey was cut, below the Camujiro Fm. (pre-Camujiro layers), a cut-rich in sediments and fossil association from Aptian-Albian, possibly coeval with Provincial Fm. Central Cuba.

Radiometric data show 95 ± 5 to 64 ± 5 million years ages (Cenomanian-Danian) for the various granitoids E of Camagüey (Marí-Morales 1997), while the age of granitoids in the whole east-central Cuba comprises a somewhat different range, Aptian to Campanian. In both cases, this means that the intrusives, together with the volcanics, form a large volcanic-plutonic complex, active for about 40–60 million years.

Some reports, in the 1980s, assigned thicknesses above 10 km to volcanic-sedimentary successions in the territory between Ciego de Ávila and Camagüey (Belmustakov et al. 1981), but this seems exaggerated. The absence of a clear linearity and orientation in most outcrops point to a lying nearly horizontal for these layers in this broad region.

Further east, in the mountains of Maniabón, northwest of Holguin, some features of Volcanic Arcs Terrain change. For its composition, two mélanges (Iberia and Loma Blanca) are distinguished. The so-called "Iberia Fm." contains lavas and pyroclastic of composition between andesites and basalts and occupies much of the area. The "Loma Blanca Fm." has a generally more acidic and more varied composition and tuffs often find zeolitised. It emerges toward the western portion of Alturas de Maniabón. In both units, there are many bodies of serpentinites, tectonically located (profile 13–14, Fig. 2.6; Kozary 1968; Knipper and Cabrera 1974) that mix with the volcanic-sedimentary rocks, forming a mélange. At las Alturas de Maniabón granitoids are absent, in contrast to its abundance in the territory of Ciego de Avila–Camagüey–Las Tunas. This may relate to the fact that Maniabón is located north of the extension to the East of the granitoid belt.

The top of the KVAT contains sedimentary beds of Campanian to Maastrichtian age. At the west of the Yabre lineament, the cover contains accumulated deposits in

Fig. 2.6 Outcropping of serpentinitic-volcanic melánge near Guardalavaca, Holguin

basins that received hundreds of meters of thickness of volcanic turbidites between the late Campanian and the Maastrichtian (Vía Blanca Fm., Brönnimann and Rigassi 1963; Piotrowski 1987; Gil-González et al. 2007), coming from nearby elevated areas. On these rest detrital-carbonated deposits (Peñalver Fm.), linked to the of the Cretaceous/Paleogene boundary event (Takayama et al. 2000; Tada et al. 2002; Goto et al. 2008). In the central and Camagüey provinces KVAT cover contains terrigenous and carbonate clastic sediments in its lower portion (Monos, Duran, and others formations), with some interbedded pyroclastic in Central Cuba (Kantchev et al. 1978), from local focus, possibly located south of Santa Clara and the deposits of the K/Pg boundary. The younger layers are shallow water carbonate deposits (Yáquimo, Cantabria, and other formations).

The contact Upper Cretaceous cover with KVAT underlying successions has a variable nature. In the northwest of Artemisa (Bahia Honda-Mariel), the cover rests with structural unconformity (Campanian unconformity) on KVAT and ophiolitic belt. In the provinces of Havana and Mayabeque, contact appears to mark a takeoff tectonic plane (decollement) on the KVAT substrate very deformed and ophiolitic melánge. The decollement possibly developed taking advantage of the Campanian unconformity. Between Varadero lineament and Yabre lineament, contact is stratigraphic with KVAT rocks, but tectonic with ophiolitic rocks. Accepting this interpretation, it can be assumed that part of the cover remained in situ, during the orogenic event and part participated in local decollement.

Toward Central Cuba, from the Yabre lineament to Camagüey lineament, the contact is always stratigraphic, with an outstanding unconformity of terminal Campanian age. In the extended area from La Trocha fault to the west of Las Tunas, basal units are terrigenous and of Upper Campanian age (Duran Fm.). The rocks of the arc cover rest on both the granitoids and on the different lithostratigraphic units of the KVAT, evidencing a structural discordance that must be correlated with that of the late Campanian more to the west. In the Alturas de Maniabón, relationships KVAT—Campanian–Maastrichtian cover are complex. On the one hand, shallow carbonated deposits (Tinajita Fm., Figure 2.7) always have tectonic contacts and are part of a large regional mélange, but terrigenous units (La Jíquima and Sirvén

Fig. 2.7 Silla de Gibara. The top of this remarkable elevation (and many other hummocks in the Alturas de Maniabón) is formed by carbonated bank limestones (Tinajita Formation), resting tectonically on serpentinite

formations) possibly lie stratigraphically on the mélange. This points to a violent tectonic event during the Maastrichtian.

In tectonic schemes of the 1950s to 1980s of last century, the ophiolites, along with rocks of the Volcanic Arcs Terrain and piggyback basins as well as other units, were often grouped in a large tectonic unit called "Zaza zone", or Cuban eugeosinclinal (Kozary 1968; Knipper and Cabrera 1974; also see Linares Cala et al. 2011). With the decline of the geosynclinal theory and use of the ideas of plate tectonics in the past 40 years, the term eugeosinclinal has been abandoned, although it occasionally appears. The same applies to the sections of the northern Mesozoic passive paleomargin, then known as miogeosinclinal.

2.2.2.4 Southern Metamorphic Terrains

Outcrop in Isla de la Juventud and the Guamuhaya (Escambray) massifs. In several aspects are one of the most enigmatic and, at the same time, complex elements of Cuban geology.

The lithostratigraphy of Guamuhaya (Escambray) has points of contact and divergence with the present in the Isla de la Juventud. Like this, the lower sections have a siliciclastic protolith, while the top is made of carbonate protolith (Fig. 2.8), in whose lower part there are Upper Jurassic fossils. The degree of metamorphism of these rocks varies from greenschist to high-pressure metamorphites (Millán Trujillo 1997a, profile 7–8, Fig. 2.3). In the older layers, there are some lenses of metavolcanic green schists. Above, the rocks differ. Although still dominating the marbles, there are units where the metavolcanics content varies from significant (Los Cedros Fm.) to dominant (Yaguanabo Fm.). A few bad preserved fossil remains indicate an age between the Tithonian and Albian to the basal portion of this upper part of the cut. Younger rocks are considered Cretaceous (Millán Trujillo 1997a).

The Guamuhaya massif forms two large dome-shaped structures (Trinidad and Sancti Spiritus), well marked in relief, separated by the Trinidad basin, to S, and to N, by Eocene clastic deposits of Meyer Fm. The domes are clearly reflected in the

Fig. 2.8 On the *left*, deposits of possible Jurassic age (Cobrito Formation) at Guamuhaya terrain (Escambray). On the *right*, marbles with the folded foliation, also possible Jurassic

package of Cretaceous volcanic-sedimentary rocks that are arranged in tectonic contact on the metamorphic complex. This deviation of the structures is perceptible in the tectonic map at distances greater than 20 km from the edge of metamorphic Guamuhaya Massif, indicating that the domes are deep structures (profile 7–8, Fig. 2.3). The age of the KVAT emplacement on the Guamuhaya massif is Late Cretaceous since a basal slip plane (decollement) affects arc rocks of the Upper Cretaceous, but apparently not to its Campanian–Maastrichtian sedimentary cover. Millán Trujillo (1997b) says that this event is simultaneous with the metamorphism of greenschist facies recorded throughout Guamuhaya, which was estimated as about 85 million years old (Santonian). In the author's opinion, a slightly younger age (Campanian) is most likely.

Millán Trujillo (1997a) distinguishes four major tectonic packages (thrusts) called major tectonic units (MTU, Fig. 2.3). The lowest in the structure is the first MTU and the fourth MTU is the highest. Internally, each MTU (except the fourth) is divided into smaller nappes. The first and second MTU in the tectonic map are distinguished as lower MTU (MTL) and the third and fourth are higher MTU (MTH). The massif has a high-pressure metamorphism, with an inverted zonality between the first and third MTU. The metamorphic peak is reached in the third unit, with records of pressure/temperature 15–23 kbar/470–630 °C (Gavilanes nappe by Stanek et al. 2006). The fourth MTU contains metamorphosed rocks under high pressures, but lower temperatures than the third. A subsequent episode of metamorphism of greenschist facies (quoted above) affects the whole massif. The first metamorphic event may correspond to approximately 106–100 Ma (Albian), according to Millán Trujillo (1997b).

Numerous serpentine and metamaphyte bodies are included tectonically, often forming serpentinite mélanges. The most notable of them constitute high-pressure amphibolites (12–14 kbar/550–580 °C, Stanek et al. 2006) of Yayabo Fm. (Millán Trujillo 1997a). These may originally be part of the base of the volcanic arc, tectonically mixed with Guamuhaya metamorphites (Stanek et al. 2006). Some of the metamaphytes of igneous protolith (especially those not linked to serpentinites) may be manifestations of continental margin magmatism (e.g., Esquistos Felicidad or metavolcanics in the top of the La Chispa Fm.), similar to that observed in the SR-APN-E unit and the Arroyo Cangre Fm.

Stanek et al. (2006) estimate that the exhumation of the Guamuhaya massif began about 70 million years ago (Maastrichtian), although the first clasts of metamorphites in sedimentary basins of its periphery appear in rocks of Middle Eocene (45 million years). The uplift process of both domes remains active today.

In our opinion, sections of metamorphosed extensional (passive) continental margin type present in the lower portion of all MTU could form a terrain (nearby), related to the terrain Pinos. In MTU 3 and 4, that terrain is geographically isolated, but in MTU 1 and 2 is in tectonic contact, interdigitated with metamorphosed volcanic-sedimentary successions (Los Cedros and Yaguanabo formations) of possible Cretaceous age, which may be originally linked to the sequences of Cretaceous volcanic arcs.

2.2.3   Paleogene Folded and Faulted Belt

The KVAT—Upper Cretaceous Cover unconformity, together with the burial of the southern metamorphic terrains under TKVA and other events are evidence of the KVAT collision with the southern paleomargin of North America. This collision caused the end of the Cretaceous volcanic arc and the restructuring of plate boundaries (Cobiella-Reguera 2000, 2008) to early Paleocene. The theme of the early and middle Paleogene tectonic zonality has been treated in Cuban geological literature (Dilla and Garcia Méndez 1984; Cobiella-Reguera 1988, 2009; Iturralde-Vinent 1996c), but not with the depth required by the subject. In the tectonic map of Cuba (Cobiella-Reguera 2016), the following main structures of the early and middle Paleogene, where sediment accumulation occurs, are recognized (Fig. 2.9):

- Foreland basin.
- Paleogene piggyback basins.
- Sierra Maestra-Cayman Ridge volcanic arc.
- Upper and middle Eocene synorogenic basin of Southeastern Cuba.

The last two structures are not described in this chapter as being beyond the scope of this publication.

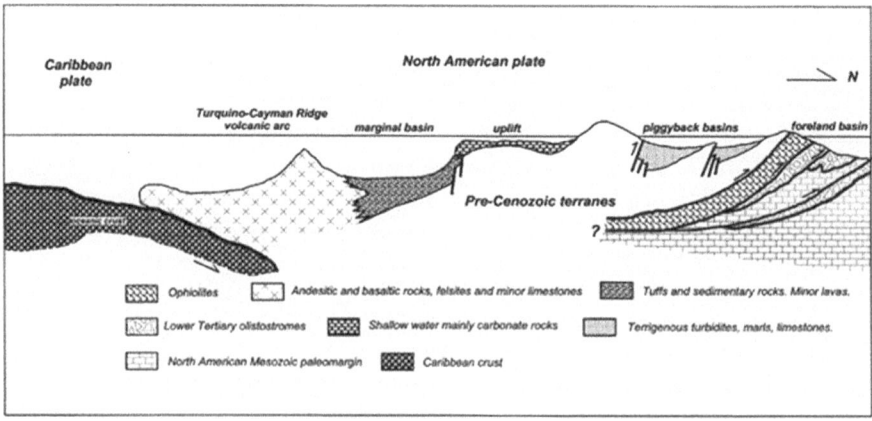

Fig. 2.9 North–South schematic tectonic profile showing the different structures present toward the center of Cuba in the range Paleocene–Middle Eocene. *Note* that in this interpretation (Cobiella-Reguera 2009), the Caribbean plate is subducted under the Paleogene volcanic arc and structures of thrust in northern Cuba, located several hundred kilometers to the north. This situation is not attributed to a collision of Bahamas platform with the Caribbean plate, but a more superficial process, linked to a thrust sheets tectonics, proper from the inner edge of foreland basins, according to Miall, in Busby and Ingersoll (1995)

2.2.3.1 Foreland Basin

Throughout North of Cuba, from Pinar del Río northwest until Gibara (Holguin), North American paleomargin rocks are covered by the foreland basin deposits (Fig. 2.9). These are successions accumulated in front of the over thrust generated during the Cuban orogeny, as a result of the erosion of its frontal region and the rapid subsidence of the basin (due to the weight of the nappes) creating the space for accommodation of large amounts of sediment. Sedimentation in these depressions is coeval with the orogenic deformation and the dating of deposits mark the event age (Busby and Ingersoll 1995). There is a strong imbrication between the tectonic scales from the southern portion of the foreland basin, formed mainly by olistostromes, and the scales of ophiolitic rocks, KVAT and the North American paleomargin. This strip of narrow interweaving of Mesozoic rocks of North American paleomargin, the ophiolitic rocks and KVAT with deposits of Paleogene foreland basin (Hatten 1957; Pardo 1975; Piotrowski 1987; Pszczółkowski 1994a; Iturralde-Vinent 1997; Cobiella-Reguera 2008) is caused by a combination of compressional and gravitational tectonics (Fig. 2.9). It has been called "Paleocene–Eocene deformed belt" (Cobiella-Reguera 2000) or "Northern Cuba folded and faulted belt". Some geologists (Rojas-Agramonte et al. 2006; Pindell et al. 2006) have interpreted this scaled belt as an accretionary prism, linked to a subduction zone located to the north of Cuba but, taking into account the simultaneity of the folded belt deformations with the activity of Paleogene volcanic arc (Fig. 2.9), there are arguments for another possible model.

The foreland basin modifies some of its features along the strike. From the composition, age and degree of deformation of their deposits and architecture of depression, we can distinguish the following sectors:

1. West Sector. It comprises from northwest Pinar del Rio to the Yabre lineament. The deposits of the foreland basin in this sector are known as Manacas Fm., up to around Havana (Martin Mesa structure) and, from this city to the east, are designated as Vega Alta Formation (Fig. 2.10). Manacas-Vega Alta rocks lie as huge

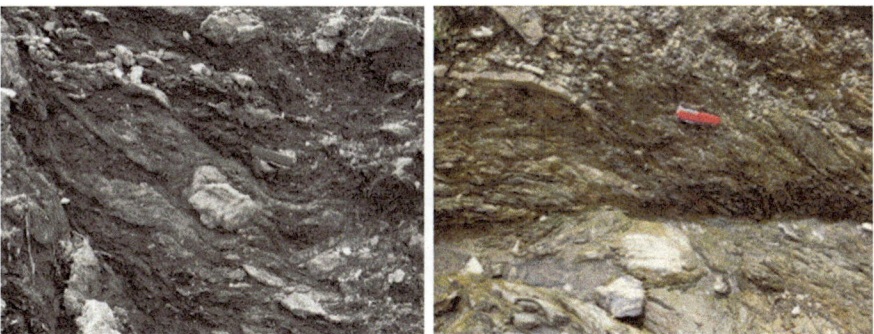

Fig. 2.10 Vega Alta Formation, strongly tectonized chaotic succession (observe the foliation bordering clasts) that characterizes the inner portion of the foreland basin in North Central Cuba

tectonic lenses sandwiched between tectonic scales of Mesozoic rocks of North American paleomargin. Its thickness is in the order of tens to hundreds of meters. They are mostly chaotic deposits, composed of clasts of different Mesozoic units (except the southern metamorphic terrains) greatly affected by intense deformation immediately after its accumulation (Pszczółkowski 1994a, b; Bralower and Iturralde-Vinent 1997; Cobiella-Reguera 1998). Genetically, are gravitational deposits (olistostromes) formed in front of thrust sheets.

2. Central Cuba Sector. It covers a band extending from the Yabre lineament to Tamarindo (Ciego de Avila), in the La Trocha fault zone. In its southern portion (inner basin) there are crushed, chaotic, olistostromic deposits, rich in ophiolite clasts and rocks of Placetas unit with minor amounts of volcanics from Cretaceous arcs (Formation Vega Alta, Fig. 2.10). Although, obviously the Vega Alta Formation rocks were originally deposited unconformably on the Mesozoic paleomargin, their current contact with paleomargin deposits is probably tectonic in the vast majority of cases. In essence, the layers of the foreland basin are divided into several tectonic scales that, in depth, possibly join in a main basal decollement plane.

Deposits located in the northern half (external foreland basin), which lie on Camajuaní and Remedios units, mostly contain clasts derived from these units (Iturralde-Vinent et al. 2008, Fig. 2.11). Its lower contact is considered tectonic for the Vega Fm. (Paleocene–Middle low Eocene), which rests essentially on the Camajuaní unit, and stratigraphic, when it comes of formations Grande (Paleocene–Lower Eocene) and Caibarién (Lower Eocene–Middle low Eocene), which lie on the Remedios tectonostratigrafic unit (Profile 7–8, Fig. 2.3).

3. Ciego de Ávila-Camagüey Sector. It is located between La Trocha and Camagüey lineaments, north of these provinces, with discontinuous outcrops between near Bolivia (Ciego de Avila) to the north of Minas (Camagüey). As in

Fig. 2.11 Carbonated brechya of Vega Formation near Camajuaní. These deposits are the main litho stratigraphic unit distinguished in external foreland basin of the North Central Cuba

Central Cuba, a zonality occurs in the composition of the sedimentary infill, which allows distinguishing two depressions. The inner basin contains thick clastic deposits, largely olistostromes, of the Senado Formation, very similar to the Vega Alta Formation, with abundant clasts of ophiolitic origin. Some of the olistolites present can reach 1 km in diameter. There are also clasts derived from Placetas North American paleomargin unit. In the matrix some sub-rounded clasts, possibly derived from KVAT, appear. According to Iturralde-Vinent et al. (2008), the formation is upper Middle Eocene to lower Upper Eocene. The external basin is composed of clastic carbonate sediments, largely derived from erosion of Cretaceous carbonate North American paleomargin deposits (Paso Abierto, Embarcadero, Lesca, Calciruditas Féliz and Venero formations, Lower and Middle Eocene). Partially overlie the Remedios unit, but northward the substrate is formed by rocks of Cayo Coco unit. It is significant the presence of fine interbedded thin tuffs in layers of Middle Eocene (Lesca Formation) showing the distant presence of a coeval volcanic source (Sierra Maestra-Cresta Caimán volcanic arc).

In the southern edge, deposits of the Senado Formation tectonically rest on external foreland basin formations, but to the north, formation lies with stratigraphic contact on the layers of the outer basin. This may be related to progress toward the north of nappes located to the south of the foreland basin, accompanied by displacement in the same direction, of the depositional front.

4. Holguin Sector. In this territory, the foreland basin has similarities to Ciego de Ávila-Camagüey, with a sedimentary infill, which also lies on the Remedios unit (Fig. 2.5). Southward, a tectonic contact separates it from the ophiolitic rocks and Iberia mélange. The equivalent of the Senado Formation is the Rancho Bravo Formation. Like the first, the Rancho Bravo Formation is largely olistostromic, with abundant ophiolitic clasts and KVAT rocks. The Embarcadero and El Recreo formations form the filling, mostly carbonate-detrital, of external basin. As partially occur in Camagüey, deposits of the inner basin (with clasts of volcanics and ophiolites) lie stratigraphically above the beds of the outer basin (clastic carbonates), suggesting a shift to north of the thrust mantles front contacting with Rancho Bravo Formation.

In fact, despite its limited dimensions, the situation of the foreland basin in Holguin is fundamental to understand the relationships between the folded and faulted belt north of Cuba and the Paleogene volcanic arc, located further south. Relations outlined in Fig. 2.12 summarize a regional interpretation that responds to the model of a back-arc collisional foreland basin.

In the four studied foreland basin regions, on their rocks generally lie ophiolite association cuts, more rarely of the KVAT, tectonically emplaced, or from the cover.

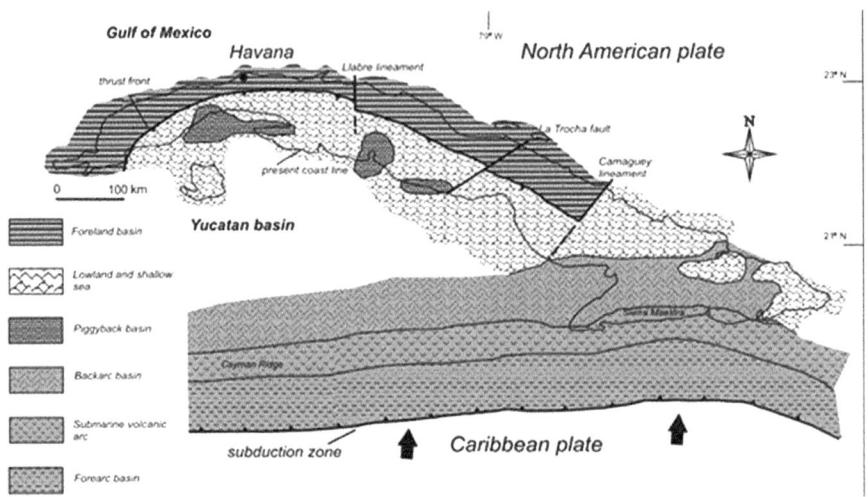

Fig. 2.12 Paleogeographic map of Later Paleocene–Early to Middle Eocene interval in Cuba and southwestern part of the Caribbean (Cobiella-Reguera 2009)

2.2.3.2 Piggyback Basins

Piggyback basins (PB) are small depressions, developed on the back of thrust sheets during the advancement of these (Busby and Ingersoll 1995). In the Cuban territory, there is evidence of the development of several of these basins, especially during the early Paleogene (Cobiella-Reguera 2009; Linares Cala et al. 2011). As is the case with other structures, Cuban piggyback basins modify some of its features from one region to another.

Piggyback basins in Central Cuba reach a development much higher than their counterparts in the West (west Yabre lineament). Three types are distinguished: Cienfuegos and Santa Clara (Fig. 2.3) in the west of the territory and, Cabaiguán, the east. In all is visible the participation of cover rocks of volcanic arcs terrain, only slightly older (and concordant) in the substrate, so that its development is evident in inherited depocentres from basins that subsidize since the late Cretaceous. Therefore, in the tectonic map they differ as inherited piggyback basins. In these are recorded some interbeds of reworked tephra and tuffs of late Maastrichtian and Paleocene age, evidence of a weak explosive volcanic activity in nearby areas. A generation of younger pyroclastic (Lower and Middle Eocene) is presented geographically limited to the southeastern Cabaiguán basin (Bijabo Formation of Kantchev et al. 1978). The Eocene intercalations should be linked to the volcanic foci of submarine Sierra Maestra-Cresta de Caimán volcanic arc (Cobiella-Reguera 2009, Fig. 2.12).

In Camagüey, the clastic composition and age of the filling of piggyback basins have particular features. Piggyback basins of Ciego de Avila and Camagüey have two varieties. To the north and mostly resting on ophiolitic rocks, lie thick clastic sediments, assigned to the Taguasco Formation (Upper Paleocene–Lower Eocene). By its composition, the clastic material obviously comes from KVAT erosion. At south, the filling of the basins is younger (Lower and Middle Eocene) and more varied, with sandstones, siltstones, carbonate intercalations and some layers of tuff and tuffites (Vertientes Formation) and shallow carbonate facies (Florida Formation; Profile 9–10, Fig. 2.4). The coeval volcanic material must come from Sierra Maestra-Cresta de Caimán arc (SMCVA, Turquino arc, sensu Cobiella Reguera 1988; see also Bresznyansky and Iturralde-Vinent 1978), which should be located relatively close to the south. From the methodological point of view, the dilemma of how to classify the deposits of Vertientes Formation arises, as also they admit to being considered part of the northern edge of the SMCVA back-arc basin. In the tectonic map, it was considered Vertientes Formation southern outcrops, in the south of Camagüey, as the filling of an intermediate depression between a piggyback basin and a back-arc basin (BA-PBPg).

In Holguin, southern the Alturas de Maniabón, the Paleogene siliciclastic-carbonate deposits from the Upper Paleocene and Middle Eocene contain some abundant interbedded pyroclastic and, even some isolated intrusive bodies (Paleogene?) are presented, that apparently cut structures originated during Maastrichtian deformations. Upper Paleocene deposits are thick terrigenous (Haticos Fm., Jakus 1983), with some resemblance to the Taguasco Fm. of Camagüey, but derived here from erosion of nearby ophiolitic massifs. The younger layers contain finer siliciclastic rocks and alodiapics limestone beds, in both cases with frequent turbidite features (Vigia Fm., Jakus 1983). The Haticos and Vigia formations contain interbedded pyroclastic. For this reason, and their spatial relationship with the foreland basin in the region of Gibara, in the tectonic map appear as a particular part of the north flank of back-arc basin (BA-PBPg) of SMCVA.

2.2.3.3 Synorogenic Basins and Oil–Gas Potential in Cuba

According to Cruz Orosa (2012), the formation of major Cuban tectonic corridors was coeval with the orogeny, which led to the segmentation of the orogen in a series of structural blocks that evolved independently. Linked to one of these tectonic corridors is the Central Cuba synorogenic basin (Central Basin), related to the La Trocha fault zone. The kinematics of the plates and the structural evolution of La Trocha fault zone indicate that the Central Basin is a polygenic tear basin and that the formation of this system (i.e., fault zone—tear basin) was a consequence of the oblique collision that occurred during the Paleogene between the Caribbean Volcanic Arc and the margin of the Bahamas (North American plate). In the opinion of this author, from tectonostratigraphic analysis of synorogenic basins, and its structural and tectonic implications, can be established a set of criteria for hydrocarbon exploration in Cuba. In this sense, the author states that the characteristics of Cuban petroleum systems are tightly controlled by the structure of the orogen. Therefore, there have been identified three main plays systems, which are associated to the Cuban folded belt, major tear structures and foreland system, respectively. The author suggests that the deposits discovered in the plays systems associated with the folded belt and tear structures may contain crudes of any quality depending on the primary characteristics and maturity of the source rock, the type and magnitude of migration, the overlapping or not of different oil systems and/or the occurrence of secondary processes. It also believes that these deposits are mostly small in volume of its reserves and will be linked to structural traps type: duplex, triangular areas and back thrusts, in the plays system of the folded belt and; faulted anticlines, flower structures and seals against faults, in the plays system associated with tear structures. Instead, he suggests that undiscovered oil fields in the foreland plays system may have high-quality crudes and large volumes of reserves. However, it states that one must consider that, although the play system associated with the foreland is currently the greatest interest attracts by their assessment of risk/reward; by geochemical and structural characteristics of Cuban Orogen, other areas should not be dismissed.

2.2.4 Eocene-Quaternary Cover

The cover (Neoauthocton sensu Iturralde-Vinent 1997) comprises the large upper structural stage of the Cuban orogen. From Western Cuba to northwestern Holguin, it includes little deformed strata, accumulated after the Cuban orogeny. In eastern Cuba, south and east of the Guacanayabo–Bahía de Nipe lineament, the cover embraces the layers accumulated after the Eocene magmatic activity in the late Eocene. Since the events of the Cuban orogeny not concluded simultaneously throughout the territory are affected, the chronostratigraphic fingerboard of the cover is different in different regions. There is strong evidence of the decisive role played by several narrow strips arranged transversely to the general direction of the structures generated by the Cuban orogeny. These structures (transverse tectonic lineaments) have a linear character, better defined in certain sectors, less clear in other. In the tectonic map of Cuba (Cobiella-Reguera 2016), the following tectonic lineaments related to the chronology of the deformations occurred during the Cuban orogeny are identified: 1—Yabre lineament, 2—La Trocha lineament, 3—Camagüey lineament, 4—Guacanayabo-Bahía de Nipe lineament.

As previously noted, the orogenic events are genetically linked to the arrival in the foreland basin of chaotic deposits (synorogenic), whose age is dated acceptably well. West of Yabre lineament are Manacas and Vega Alta formations (Lower Paleocene–Lower Eocene). Between Yabre and La Trocha lineaments correspond to the Vega Alta Formation (Lower Paleocene–Middle low Eocene). East of La Trocha lineament, the Senado Formation (Middle Eocene) spans throughout the foreland basin to the north of Camagüey and, further east, the Rancho Bravo Formation (Middle Eocene) contact tectonically with ophiolitic rocks west of Gibara.

The above data, combined with other evidence, allow concluding that west to the Yabre lineament, orogenic processes elapsed between the Early Paleocene and

Early Eocene. Between Yabre and La Trocha lineaments, orogenic deformation conclude up to early Middle Eocene and, from La Trocha lineament to the northwest of Holguin, the orogeny takes place during the late Middle Eocene and possibly reaches the Late Eocene. Therefore, the base of the cover is markedly diachronic.

The cover outcrops in more than 50% of the Cuban territory. As previously noted, it is not homogeneous either horizontally or vertically. From the study of the cover's sediment thickness in 40 wells, its average thickness is approximately 760 m, with a significantly greater thickness reported in wells Candelaria-1 (about 3740 m) located near Pinar fault, Las Mangas-1 (2145 m) and Vegas-1 (2500 m) in southern Mayabeque.

According to the data available and considering the limitations of the scale, it is proposed to distinguish four substages arranged in the following sequence (from bottom to top): A, B, C, D. Together, these substages range from the Lower Eocene up to Quaternary for the territory west of the Guacanayabo- Bahía de Nipe lineament. To the southeast of Cuba, east of the above lineament, in the substrate over thrust deformations are not recorded and there are some differences in the chronostratigraphic fingerboard of the cover respect to the cut in the west and center. Here, the following substages are proposed: A', C', and D', covering the Upper Eocene to Quaternary.

Substage A. Transitional successions

In several regions, it is included in the basal portion of the cover successions with some degree of structural complication and evidence of accumulation in even unstable conditions. Transitional successions (A) are located in areas at the base of the cover. They are siliciclastic nature, largely, sandstones and immature conglomerates. They were deposited in basins of limited extent in conditions of tectonic instability, reflected by numerous synsedimentary deformations (especially submarine landslide folds). Among the formations belonging to this group are Capdevila (Lower low Eocene), extended from Pinar del Rio to Mayabeque (Brönnimann and Rigassi 1963), Marroquí and Arroyo Blanco (Kantchev et al. 1978), in the northwest of Ciego de Avila. These units contribute to dating the end of the thrusts in the Western and Central Cuba, showing they become younger from west to east.

In Matanzas, substage seems to be represented by the Perla Formation (Lower and Middle Eocene) of siliciclastic nature at its lower portion, transitioning vertically to carbonate deposits. This unit lies both on the North American paleomargin as on KVAT rocks. Therefore, the orogeny was concluded by mid-Early Eocene in the current central Matanzas. To the east, in the Trinidad basin, Meyer Fm. (Middle Eocene) is considered as the base of the cover. This unit is composed by massive brechya of metamorphic clasts, with a maximum thickness of 300 m. The Condado Formation (Subfloor B) unconformably covers the Meyer Formation.

Another structure where the substage is recorded in the Central Basin (located between the cities of Sancti Spiritus and Morón). In it, substage A is represented by the Arroyo Blanco (100–600 m thick) and Marroquí formations, both from Middle

Eocene high–Upper Eocene, which lie unconformably on structural units of the Middle Eocene (Kantchev et al. 1978).

Substage B. Western Cuba—Camagüey (up to about 77° 30' W)

The substage B is fully formed by marine deposits. At them, siliciclastics play a subordinate role with marl and limestones as main lithologies. It is mainly located in west of La Trocha lineament, with minor inliers in Camagüey. The stratigraphic record includes deposits between the Lower or Middle Eocene and Upper Oligocene. In Western Cuba, its layers lie with moderate unconformity on rocks of the transitional sequence. To the east of Varadero lineament, virtually disappear outcrops of substage A until Central Cuba, and then, B rests on the basement, with marked structural unconformity, as consequence of the late Cretaceous events and the Cuban orogeny.

Between the Varadero and Yabre lineaments, substage B outcrops are located mostly toward the center of the territory, being part of the core of large antiforms. Sedimentary thicknesses appear to be discrete. From the available data, a thickness not exceeding 700–800 m can be considered (Brönnimann and Rigassi 1963).

Stratigraphic successions of substage B (Eocene–Upper Oligocene) between Mayabeque, Cienfuegos and Villa Clara largely are also formed by marl and limestone, with fine terrigenous deposits, all of marine origin (Nazareno, Hatillo, Peñón, Jicotea, La Jía, Caunao, Tinguaro and Saladito formations). All have a total thickness of the order 400–500 m in the structures between Bejucal and Cidra and, about 1000 m, in Cienfuegos basin. In these regions, the subfloor involved in the core of various structures of anticline type, includes Bejucal, Madruga, Cidra, Coliseo, Cantel-Camarioca, Perico-Colón and the broad Santo Domingo syncline. This, coupled with the frequent presence of subfloor deposits in deep drilling, shows its extensive underground distribution.

South of the Guamuhaya massif (Trinidad basin), the substage B shows a different lithological character. On the substage A and/or the basement rest, unconformably, about 1000 m of siliciclastic deposits with southern homoclinal lying (Condado Formation Upper Eocene–Oligocene). The cut ends with carbonated layers (Las Cuevas Formation, Oligocene). The equivalent of the Condado Formation in Cienfuegos basin, located to the west, is the Damují Formation. In both sequences, the siliciclastic grains come from the erosion of Guamuhaya Mesozoic metamorphites. The maximum thickness of the substage B in Trinidad basin should exceed 1200 m.

The Central Basin is a poorly defined geological structure in Cuban literature, including therein, successions from both the basement and the cover. In the tectonic map, it is considered that the depression is a structure of the Eocene–Quaternary cover that should include only spatially and genetically (?) deposits associated to La Trocha lineament. In the Central Basin, the substage B is represented by Chambas, Tamarindo, and Jatibonico formations, covering the Oligocene. The first two are carbonates, but the third contains abundant terrigenous material. Substage B in the

Central Basin includes deposits of Lagunitas Fm., reaching the Lower Miocene. Its terrigenous nature, with abundant clastic material derived from metamorphic rocks, approaches it to substage B characteristic units near the Guamuhaya massif. The fact of reach the Lower Miocene must be connected with the activity associated with La Trocha lineament. Tectonic map (Cobiella-Reguera 2016) clearly shows the spatial continuity of the outcrops of the Central Basin and the Trinidad basin, pointing to a genetic connection.

Further east of the Central Basin, substage B outcrops occupy limited areas. Eastward, they correspond to Nuevitas Formation, which lies with a marked structural unconformity on various units of the base. The unit, of about 50 m thick, is composed of marl and limestone interbedded with gypsum, resting as little deformed strata. East of the city of Camagüey, belong to this substage the strata of Saramaguacán Formation (upper Eocene) and Guaicanamar Formation (north of Santa Cruz del Sur).

Substage C. Western Cuba—Holguin northwest

Substage C strata (Upper Oligocene–Middle Miocene) lie separated by a discrete structural unconformity from underlain subfloor B and contain the most extensive cover outcrops. Unlike A and B, outcropping mainly in inliers, C rocks occupy two large areas, a Western, from 80° 15′ W to the outskirts of the city of Pinar del Rio. In the eastern half of Cuba, the substage C comprises much of the territory between the La Trocha lineament and the Nipe bay.

In Mayabeque, the thickness substage C reaches about 600 m. In Matanzas the substage C registers between 600 and 780 m thick. In the central provinces, C is almost absent. East of La Trocha lineament, B floor strata are scarce and C layers generally lie with marked unconformity on the basement (especially KVAT and ophiolite association). In the territory between Ciego de Ávila and northwest of Holguin, the substage C again contains abundant shallow carbonate deposits (Güines and Camazán formations), although siliciclastic deposits (Pedernales and Los Arabos formations) and carbonate-siliciclastic (Vazquez and Paso Real formations), accumulated in shallow marine and coastal conditions, become relatively more frequent.

South of the city of Holguin substage C reaches between 400 and 800 m of thickness, in north Las Tunas more than 200 m (Jakus 1983).

Substage D

Unlike underlying substages, the D (Upper Miocene-Quaternary) contains abundant inland, coastal and shallow marine sediments. Generally, they are arranged away from the axial region of Cuba, but there are some inland areas where they reach a certain extension (for example, in the province of Ciego de Ávila and the Cauto river basin).

In Vegas and Broa basins (South Mayabeque), the occupied strip by D is greatly reduced. In this section, the substage contains only marshy quaternary sediments.

In the Cienaga de Zapata and the Cienfuegos basin, on the surface are mainly peat that lie on coastal formations (Vedado and Jaimanitas) or fluvial (?—Villaroja) which, in turn, rest discordant on the substage C, except in the eastern edge of the Cienfuegos basin, where the substage is located on the basement or substage B.

Deep Cochinos bay is the geomorphological reflection of the homonym graben, very young structure in the NNW direction.

In the Trinidad basin, substage D basal deposits are formed by shallow carbonate of Vedado Fm. that, apparently, lie in a small angular unconformity on substages B and C. Further south are the deltas of the Agabama and Zaza rivers, which join the East with the South Jatibonico. This low coast extends to Júcaro, forming a rela- tively extensive coastal plain in the south of the province of Sancti Spiritus, tes- timony of subsidence and Quaternary progradation in South Central Basin. In the coastal plain of Southern Camagüey, the substage forms a thin layer of Quaternary deposits, currently suffering erosion.

Along the northern coast of Western Cuba, width of D outcrops is very small, predominantly of marshy deposits until Bahia Honda. From this town to Matanzas bay, the substage D forms a narrow strip mostly represented by coastal Quaternary carbonates, modeled on marine terraces, whose maximum height is presented on its eastern edge, at the entrance to the Matanzas bay. Geomorphology of Matanzas bay and the distribution of D deposits show the existence of an active syncline along the valley of the Yumurí river. Moreover, the marine terraces on the western side of the bay and the west coast are a classic periclinal close sinking to the east.

In Villa Clara northeast and Sancti Spiritus small enclaves of subfloor C rocks (Güines Formation) are recorded in areas occupied by D. Currently, some tiny rocky islets, formed by rocks of this formation are presented, which sharply contrast to the "normal" keys, formed by quaternary deposits. Two evolutionary models seem feasible to explain the rocky keys: 1—these rocky keys (Salinas, Guainabo and others) east of Caibarién are the tops of small elevations, possibly buried because of Holocene rise of sea level. The substage C enclaves can be explained by a more advanced stage of this transgression and; 2—the keys might be related to upward movements of the Jurassic diapirs thus, immediately to the east, the three major salt diapirs Central Cuba are located: Punta Alegre, Turiguano and Cunagua. These structures are alive and are reflected in the current peculiar relief, determined by geographical features such as Laguna de Leche, the Cunagua hill and the river network south of this elevation. Therefore, there should not rule out the possibility that the above "rocky keys" may be linked to the current upward movement of the Jurassic salts, which, in turn, could be related to tectonic activity according to La Trocha lineament.

East of Cunagua hill, the D is mostly occupied by thin marshy deposits, lying on C, except in the region of Nuevitas, where they rest on the subfloor B. Nuevitas bay is a small depocentre, and supplied by the Saramaguacán and other rivers. Eastward, the strip of deposits D is mostly coastal biogenic carbonates (Jaimanitas Fm.). From Gibara, the basement of D is formed by calcareous and loamy marine deposits of the Upper Miocene-Pliocene.

2.3 Conclusions

The Cuban orogen can be divided into two major structural and stratigraphic units: basement and cover. The basement is the mega complex of igneous, metamorphic, and sedimentary rocks that lies below the little deformed cut of the cover. It is divided into several tectonic larger units, according to its structural style and age of the rocks. We can distinguish three large basement complexes: (a) Proterozoic basement, (b) Mesozoic basement, and (c) Paleogene folded belt.

The Proterozoic basement emerges are very limited.

The Mesozoic basement consists of four complexes of very different nature: the Mesozoic paleomargin of the SE North American plate, which presents Jurassic-Cretaceous sequences with varying degrees of deformation; the remaining three, the ophiolite association, the Cretaceous volcanogenic successions and southern metamorphic terrains, have traits of tectonostratigraphic terrains.

The links between the four regional structures of the Paleogene folded belt are much clearer and, notwithstanding the considerable deformations and horizontal transport suffered by some, the primary spatial relationships (paleogeography) between them is essentially preserved. At the Paleogene folded and faulted belt are distinguished:

- Foreland basin successions.
- Piggyback successions.
- Cuts of Sierra Maestra-Cresta Caimán volcanic arc.
- Synorogenic basin of Middle and Upper Eocene of South Eastern Cuba.

The cover, composed of deposits of Lower or Middle Eocene to Quaternary age, comprises the younger deposits of the stratigraphic cut, little dislocated in relation to the underlying layers, usually separated of these by remarkable structural discordances. Across Cuba, the cover is totally devoid of evidence of magmatic, metamorphic, and hydrothermal activity.

References

Andó J, Harangi S, Szakmany B, Dosztaly L (1996) Petrología de la asociación ofiolítica de Holguín. In: Iturralde-Vinent M (ed) Ofiolitas y arcos volcánicos de Cuba IUGS/UNESCO. Miami, International Geological Correlation Program, Project 364, Contribution No 1, pp 154–176

Belmustakov B et al (1981) Informe del levantamiento geológico de las provincias Ciego de Ávila, Camagüey y Oeste de Las Tunas, escala 1:250 000. Instituto de Geología y Paleontología de la Academia de Ciencias de Cuba. Inédito. Archivo del Instituto de Geología y Paleontología, La Habana

Blein O, Guillot S, Lapierre H, Mercier de Lepinay B, Lardeaux JM, Millan-Trujillo G, Campos M, García A (2003) Geochemistry of the Mabujina Complex, Central Cuba: implications on the Cuban cretaceous arc rocks. J Geol 111:89–101

Bralower T, Iturralde-Vinent MA (1997) Micropaleontological dating of the collision between the North America and Caribbean plates in western Cuba. Palaios 12:133–150

Bresznyansky K, Iturralde-Vinent MA (1978) Paleogeografía del Paleógeno de Cuba oriental. Geologie en Mijnbow 57(2):123–133

Brönnimann P, Rigassi D (1963) Contribution to the geology and paleontology of the area of the city of La Habana, Cuba and its surroundings. Eclogae Geologia Helvetiae 56(1):193–430

Busby C, Ingersoll R (eds) (1995) Tectonics of sedimentary basins. Editorial Blackwell Science, 579 pp

Cobiella-Reguera JL (1988) El vulcanismo paleogénico cubano. Apuntes para un nuevo enfoque. Revista Tecnológica XVIII(4):25–32

Cobiella-Reguera JL (1998) Las melánges de Sierra del Rosario, Cuba occidental. Tipos e importancia regional. Min Geol XV(2):3–9

Cobiella-Reguera JL (2000) Jurassic and Cretaceous geological history of Cuba. Int Geol Rev 42 (7):594–616

Cobiella-Reguera JL (2005) Emplacement of Cuban ophiolites. Geol Acta 3(3):273–294

Cobiella-Reguera JL (2008) Reconstrucción palinspástica del paleomargen mesozoico de América del Norte en Cuba occidental y el sudeste del Golfo de México. Implicaciones para la evolución del SE del Golfo de México. Revista Mexicana de Ciencias Geológicas 25(3):382–401

Cobiella-Reguera JL (2009) Emplacement of the northern ophiolite belt of Cuba. Implications for the Campanian-Eocene geological history of the northwestern Caribbean-SE Gulf of Mexico region. In: James K., Lorente M, Pindell J (eds) The origin and evolution of the Caribbean Plate. Geological Society of London Special Publication 328, pp 313–325

Cobiella-Reguera JL (2016) Texto explicativo del mapa tectónico de Cuba (borrador para Proyecto Mapa Mineragénico de Cuba a escala 1:250 000). Inédito. Instituto de Geología y Paleontología-Servicio Geológico de Cuba, La Habana, 58 pp

Cobiella-Reguera JL, Rodríguez-Pérez J, Campos-Dueñas M (1984) Posición de Cuba oriental en la geología del Caribe. Min Geol 2–84

Cruz Orosa I (2012) Las cuencas sinorogénicas como registro de la evolución del Orógeno cubano: implicaciones para la exploración de hidrocarburos. Universidad de Barcelona. http://hdl.handle.net/2445/35485

Dilla M, García Méndez L (1984) Estratigrafia y sedimentologia de las cuencuas superpuestas de Cuba Central. Serié Geológica, Instituto de Geológia y Paleontológia, Academia de Ciencias de Cuba 3:101–154

Draper G, Barros JA (1994) Cuba. In: Donovan KS, Jackson TA (eds) Caribbean geology: an introduction

Echevarría-Rodríguez G, Hernández-Pérez G, López-Quintero JO, López-Rivera JG, Rodríguez-Hernández R, Sánchez-Arango J, Socorro-Trujillo R, Tenreyro-Pérez R, Yparraguire-Pena J (1991) Oil and gas exploration in Cuba. J Pet Geol 14(3):259–274

Fonseca E, Castillo F, Uhanov A, Navarrete M, Correa G (1990) Geoquímica de la asociación ofiolítica de Cuba. In: Larue D, Draper G (eds) Transactions of the 12th Caribbean Geological Conference. St. Croix, U.S. V. I. Miami, Miami Geological Society, pp 51–58

Gil-González S, Díaz-Otero C, García-Delgado D (2007) Consideraciones bioestratigráficas de los sedimentos siliciclásticos en Cuba, en cuencas de piggy back del Campaniano-Maastrichtiano. VII Congreso Cubano de Geología, La Habana. ISBN 978-959-7117-16-2

Goto K, Tada R, Tajika E, Iturralde-Vinent MA, Matsui T, Yamamoto S, Nakano Y, Oji T, Kiyokawa S, García-Delgado D, Díaz-Otero C, Rojas-Consuegra R (2008) Lateral lithological and compositional variations of the Cretaceous/Tertiary deep-sea tsunami deposits in northwestern Cuba. Cretac Res 29(2):217–236

Hatten CW (1957) Geologic report on Sierra de los Órganos. Inédito. Archivo de la Oficina Nacional de Recursos Minerales, La Habana

Iturralde-Vinent MA (1996a) Introduction to Cuban geology and tectonics. In: Iturralde-Vinent MA (ed) Cuban ophiolites and volcanic arcs. IUGS/UNESCO International Geological Correlation Programme. Project 364. Geological Correlation of Ophiolites and volcanic arcs in the Circumcaribbean Realm, Miami, Florida Special Contribution (1), pp 3–35

Iturralde-Vinent MA (1996b) The ophiolites's geology of Cuba. In: Iturralde-Vinent MA (ed) Cuban ophiolites and volcanic arcs. IUGS/UNESCO International Geological Correlation Programme. Project 364. Geological Correlation of Ophiolites and volcanic arcs in the Circumcaribbean Realm, Miami, Florida Special Contribution (1), pp 83–120

Iturralde-Vinent MA (1996c) Cuba: el archipiélago volcánico Paleoceno-Eoceno Medio. In: Iturralde-Vinent MA (ed) Cuban ophiolites and volcanic arcs. IUGS/UNESCO International Geological Correlation Programme. Project 364. Geological Correlation of Ophiolites and volcanic arcs in the Circumcaribbean Realm, Miami, Florida Special Contribution (1), pp 231–246

Iturralde-Vinent MA (1997) Introducción a la geología de Cuba. In: Furrazola-Bermúdez G, Nuñez Cambra K (comp.) Estudios sobre la geología de Cuba. Centro Nacional de Información Geológica, La Habana, pp 35–68

Iturralde-Vinent MA, Roque Marrero F (1982) La falla Cubitas: su edad y desplazamientos. Rev Ciencias de la Tierra y del Espacio 4:47–70

Iturralde-Vinent MA, Díaz-Otero C, García-Casco A, van Hinsbergen D (2008) Paleogene foredeep basin deposits of north-central Cuba: a record of arc-continent collision between the Caribbean and North American Plates. Int Geol Rev 50(10):863–884

Jakus P (1983) Formaciones vulcanógeno-sedimentarias y sedimentarias de Cuba Oriental en: Contribución a la geología de Cuba Oriental. Instituto de Geología y Paleontología A. C. C. de, Editorial Científico Técnica, La Habana, pp 17–85

Kantchev I, Boyanov I, Popov N, Cabrera R, Goranov A, Iolkoev N, Kanazirski M, Stancheva M (1978) Informe Geología de la provincia de Las Villas. Resultados de las investigaciones geológicas y levantamiento geológico a escala 1:250 000 realizados durante el período 1969–1975. Instituto de Geología, Academia de Ciencias de Bulgaria; Instituto de Geología y Paleontología (IGP), Academia de Ciencias de Cuba. Inédito. Archivo del Instituto de Goelogía y Paleontología, La Habana

Khudoley C, Meyerhoff AA (1971) Paleogeography and Geological History of Greater Antilles. Geological Society of America, Boulder, CO. Memoir 129

Knipper A, Cabrera R (1974) Tectónica y geología histórica de la zona de articulaciónb entre el mio y el eugeosinclinal del cinturón hiperbasítico de Cuba. In: Contribución a la Geología de Cuba, ACC. Publicación especial (2), pp 15–77

Kozary M (1968) Ultramafic rocks in thrust zones of northwestern Oriente Province, Cuba. AAPG Bull 52(12):2298–2317

Linares Cala E, García-Delgado D, Delgado-López O, López-Rivera JG, Strazhevich V (2011) Yacimientos y manifestaciones de hidrocarburos de la República de Cuba. Centro de Investigaciones del Petróleo, La Habana, 480 pp

Marí-Morales MT (1997) Particularidades de los granitoides de Ciego - Camagüey - Las Tunas y consideraciones sobre su posición dentro del arco de islas. In: Furrazola GF, Núñez KE (eds) Estudios sobre geología de Cuba. Instituto de Geología y Paleontología, La Habana, pp 399–416

Meyerhoff AA, Hatten CW (1968) Diapiric structure in Central Cuba. Am Assoc Pet Geol Mem 8:315–357

Millán Trujillo G (1997a) Geología del macizo metamórfico Isla de la Juventud. In: Furrazola Bermúdez GF, Núñez Cambra KE (eds) Estudios sobre Geología de Cuba. Centro Nacional de Información Geológica, La Habana, pp 259–270

Millán Trujillo G (1997b) Geología del macizo metamórfico del Escambray. In: Furrazola Bermúdez GF, Núñez Cambra KE (eds) Estudios sobre Geología de Cuba. Centro Nacional de Información Geológica, La Habana, pp 271–288

Pardo G (1975) Geology of Cuba. In: Nairn AEM, Stehli FG (eds) The Ocean basins and margins, vol. 3. The Gulf of Mexico and the Caribbean, pp 553–615

Pindell JL, Kennan L, Stanek K, Maresch W, Draper G (2006) Foundations of Gulf of México and Caribbean evolution: eight controversies resolved. Geol Acta 4(1–2):303–341

Piotrowski J (1987) Nuevos datos sobre los sedimentos del Cretácico Superior tardío y el Paleógeno en la zona estructuro-facial San Diego In: Pszczółkowski A (Sc. Red.) Contribución a la Geología de la Provincia de Pinar del Río. Editorial Científico-Técnica, La Habana, pp 185–196

Proenza J, Díaz-Martínez R, Iriondo A, Marchesi C, Melgarejo J, Gervilla F, Garrido C, Rodríguez-Vega A, Lozano-Santa Cruz A, Blanco Moreno J (2006) Primitive Cretaceous island-arc volcanic rocks in eastern Cuba: the Téneme formation. Geol Acta 4(1–2):103–122

Pszczółkowski A (1978) Geosynclinal sequences of the Cordillera de Guaniguanico in western Cuba: their lithostratigraphy, facies development and paleogeography. Acta Geol Pol 28(1):1–96

Pszczółkowski A (1982) Cretaceous sediments and paleogeography in the western part of the Cuban miogeosyncline. Acta Geol Pol 32:135–161

Pszczółkowski A (1983) Tectónica del miogeosinclinal cubano en el área limítrofe de las provincias de Matanzas y Villa Clara. Revista de Ciencias de la Tierra y el Espacio 6:53–61

Pszczółkowski A (1986) Secuencia estratigráfica de Placetas en el Área limítrofe de las provincias de Matanzas y Villa Clara (Cuba). Academie Polonaise des Sciences Bulletin, Serle des Sciences de la Terre 34(1):67–79. Warsaw

Pszczółkowski A (1994a) Geological cross-sections through the Sierra del Rosario thrust belt, western Cuba. Stud Geol Pol 105:67–90

Pszczółkowski A (1994b) Lithostratigraphy of Mesozoic and Paleogene rocks of Sierra del Rosario, western Cuba. Stud Geol Pol 105:39–66

Pszczółkowski A, Myczyński R (2003) Stratigraphic constraints on the Late-Jurassic-Cretaceous paleotectonic interpretations of the Placetas Belt in Cuba. In: Bartolini C, Buffler R, Blickwede J (eds) The Circum-Gulf of Mexico and the Caribbean: hydrocarbon habitats, basin formation and plate tectonics. American Association of Petroleum Geologists Memoir 79, pp 545–581

Renne P, Mattinson IM, Hatten CW, Somin MI, Onstott TS, Millán G, Linares E (1989) 40Ar/39Ar and U/Pb evidence for Late Proterozoic (Grenville age) continental crust in North Central Cuba and regional tectonic implications. Precambr Res 42:325–341

Rojas-Agramonte Y, Kroener A, García-Casco A, Iturralde-Vinent MA, Wingate MTD, Liu D (2006) Review of Zircon ages from Cuba and their geodynamic interpretations. Asia Oceania Geosciences Society, Singapore, pp 733–734

Somin MI, Millán G (1981) Geología de los complejos metamórficos de Cuba. Editorial Nauka, Moscú, En idioma ruso, 219 pp

Stanek KP, Maresch W, Grafe F, Grevel Ch, Baumann A (2006) Structure, tectonics and metamorphic development of the Sancti Spiritus Dome (Eastern Escambray massif, central Cuba. Geol Acta 4(1–2):151–170

Stanek KP et al (2009) The geotectonic story of the northwestern branch of the Caribbean Arc: implications from structural and geochronological data of Cuba. In: James KH, Lorente MA, Pindell JL (eds) The origin and evolution of the Caribbean Plate. Geological Society, London, Special Publications, 328

Tada R, Nakano Y, Iturralde-Vinent MA, Yamamoto S, Kamata T, Tajika E, Toyoda K, Kiyokawa S, García-Delgado D, Oji T, Goto K, Takayama H, Rojas-Consuegra R, Matsui T (2002) Complex tsunami waves suggested by the Cretaceous-Tertiary boundary deposit at the Moncada section, western Cuba. In: Koeberl C, McLeod K (eds) Catastrophic events and mass extinctions: impacts and beyond. Boulder, Colorado, Geological Society of America Special Paper 356, pp 109–123

Takayama H, Tada R, Matsui T, Iturralde-Vinent MA, Oji T, Tajika E, Kiyokawa S, García-Delgado D, Okada H, Hasegawa T, Toyoda T (2000) Origin of Peñalver Formation in northwestern Cuba and its relation to K/T boundary impact event. Sediment Geol 135:295–320

Chapter 3
Tectonic-Structural Regionalization with Purposes of Hydrocarbon Exploration and Mapping of New Potential Oil–Gas Goals

Abstract A summary of the main results achieved in the tectonic-structural regionalization with hydrocarbon exploration purposes, focusing mapping of new possible oil–gas targets in the regions of Onshore Blocks 9, 23, and 17–18 is given. In some of the new places of interest (Majaguillar, Southeast Motembo, Guamutas, El Pinto, and Maniabón), there were performed reconnaissance work by a profile of *Redox Complex* (complex of unconventional geophysical–geochemical exploration techniques) with positive results. In Block 9, from the presence of a complex of indicator anomalies, three locations of greater oil–gas interest were established: Majaguillar (Placetas Tectonic-Stratigraphic Unit), Southeast Motembo (Ophiolites), and Guamutas (northern boundary of Mercedes Basin). Block 23, in the locality of El Pinto (near the Zaza Dam), is distinguished as an anomalous complex (AC) given by a local gravimetric maximum, coincident with a geomorphic maximum with the same position and dimensions. Block 17–18 is distinguished, in the locality of Maniabón (south of Puerto Padre and Nuevitas bays), as an AC given by a local gravimetric–geomorphic maximum within a regional gravimetric minimum, a minimum of K/Th ratio, and U(Ra) local anomalies at the end of the AC. Definitively, for its geological significance (associated to foreland basin), it was established as main interest of all locations, the Maniabón area.

Keywords Potential fields · Airborne gamma spectrometry · Tectonic-structural regionalization · Oil–gas onshore exploration · *Redox complex* · Geomorphic anomaly

3.1 Introduction

The contribution of potential fields and airborne gamma spectrometry to geological-structural mapping and hydrocarbon exploration in the various territories depends, considerably, on the type of geology, climate, and topography of the surveyed areas. In this sense, Cuba is a privileged place for its contrasting alpine geology and its tropical climate, which determines the presence of residual soils or

© The Author(s) 2017   39
M.E. Pardo Echarte and J.L. Cobiella Reguera, *Oil and Gas Exploration in Cuba*,
SpringerBriefs in Earth System Sciences, DOI 10.1007/978-3-319-56744-0_3

in situ weathering crusts and an eminently plain relief. In addition, Cuba has an aeromagnetic and airborne gamma spectrometric survey 1:50,000 from throughout the country and a gravimetric survey, on the same scale, 80% of it.

Within the regional geophysical research on medium-large scale (1:250,000–1:50,000) which are undertaken in Cuba at present, these are related to tectonic-structural regionalization with purposes of hydrocarbon exploration, accompanied by the mapping of new possible oil–gas objectives for the substantiation of oil exploration in different onshore blocks. The chapter considers the tectonic-structural decipherment of regions of oil–gas interest (Onshore Blocks 9, 23 and 17–18) with the establishment of new potential targets from the presence of a complex of indicator anomalies (gravimetric, airborne gamma spectrometric, and morphometric). To this end, the gravimetric, aeromagnetic, and airborne gamma spectrometry fields of the different territories are processed at 1:50,000 and 1:100,000 scales, as well as the Elevation Digital Model (EDM) 30 × 30 m.

3.2 Materials and Methods

3.2.1 Materials

For carrying out the investigation, the following were used as primary sources of information:

- Grids of gravimetric and aeromagnetic field, 1:50000 scale, and airborne gamma spectrometry (channels: It, U, Th and K), 1:100,000 scale, of the Republic of Cuba (Mondelo et al. 2011).
- Elevation Digital Model 30 × 30 m of the Republic of Cuba (https://lpdaac.usgs.gov/).
- Digital Geological Map of the Republic of Cuba at scale 1:100,000 (Colectivo de Autores 2010).
- Digital Geological Map with oil purposes of the Republic of Cuba at scale 1:250,000 (Colectivo de Autores 2007).
- Digital Map of the Hydrocarbons Shows of the Republic of Cuba at scale 1:250,000 (Colectivo de Autores 2008).
- Digital Map of the Oil Wells of the Republic of Cuba at scale 1:250000 (Colectivo de Autores 2009).

3.2.2 Methods and Techniques

Methods

A theoretical–empirical research was used by the hypothetical-deductive method where, from the numerical information of physical fields and through operations of analysis and deduction, the complex phenomenon of geology and geophysics of the study regions is decomposed in their component parts: structure–tectonic, litho-logical, and oil–gas targets, in order to define its essential features under the hypotheses.

Techniques

The Processing and Interpretation System of Geophysical–Geological Data (Oasis Montaj, version 7.01) for automated processing of geophysical–geological geo-referenced information (matrix arithmetic operations and transformations of phys-ical fields and others are performed) is used.

Processing and Interpretation

For tectonic-structural regionalization with purposes of hydrocarbon exploration and mapping of new potential oil–gas objectives, the gravimetric field was sub-mitted to the regional–residual separation. It was used as the ascendant analytic continuation (AAC) for three heights (500, 2000, and 6000 m), given by the depth order of potential oil–gas objectives and seismic survey information, and the ver-tical derivative (VD) for the establishment of local anomalies with an order less than of 500 m depth.

The aeromagnetic field (reduced to pole [RTP]) was submitted to the VD and the total horizontal derivative of the tilt derivative of the field (Verduzco et al. 2004), vital for tectonic-structural decipherment of the territory. In addition, the regional field analysis (AAC) for heights of 500 and 2000 m was made. Note that tectonic limits set by the axes of the chains of maximum total horizontal derivative of the tilt derivative of the field generally agree with the chains axes of minimum of the VD of the field (RTP). In addition, only the latter provides a much less noisy structural frame and, therefore, is more detailed and preferred. It also was made quantitative estimates of depth to magnetic targets from the tilt derivative of RTP magnetic data (Fairhead et al. 2009).

The airborne gamma spectrometry considered the analysis of each channel, the K/Th ratio and the ternary map corresponding to the three element channels (K, U, and Th).

The EDM 30 × 30 m was submitted to the regional–residual separation from the AAC to 500 m, according to the experience of lead author in this type of application.

Geophysical alignments, regional tectonic dislocations that separate different tectonic-structural elements, were drawn mainly from the VD of potential fields maps, aeromagnetic, and gravimetric; considering chains of minimum (mainly) and maximum, linearity, push-ups, and interruption of contour lines and areas of high gradient of them.

The position of the alignments and limits, outlined from geophysical maps of the VD of potential fields, is considered acceptable in comparison with the corresponding geological map. The character depth of these limits is established in general as vertical–subvertical (which could be softer with depth); generally dip to the south, at northern Cuba, from comparing its position regarding the observed and processed fields of a different nature. Tectonic boundaries for the Central Basin (CB) are considered practically vertical, under the same assumptions.

Similarly, the limits of the main depressions were plotted considering a change in the characteristics of the magnetic field VD (smoothing or flattening), coincident with regional gravitational minimum.

The establishment of new locations with possible oil–gas deposits are made after considering, for blocks of the northern region of Cuba, the position of the regional belt of gravimetrical minima (linked to Northern Cuban Thrusts Belt (NCTB) [CCNC]), to which generally overlaps the Ophiolite Belt (OB) [CO]. Then, it is taken into account the presence of an indicator AC [CA]: local gravity anomalies (maximum) within the regional minima, coinciding with minimum of K/Th ratio, local anomalies of U(Ra) on its periphery, and local geomorphic anomalies (maximum). For the CB [CC], there are indicators, local gravimetric maximum within the regional minimum of the basin, coinciding with local geomorphic anomalies (maximum). In all cases, as supplementary geological information was considered, the presence of hydrocarbons shows oil wells in the locality or its proximity and the existence of rocks that could be considered seal.

3.3 Results

3.3.1 Region of Block 9

The basic geophysical materials for the study of this region are presented in Figs. 3.1 and 3.2 (gravimetric field and RTP aeromagnetics).

In the gravimetric field shown in Fig. 3.1, it is apparent that in the boundary of north center region, an area of moderately intense-positive values is associated mainly to carbonate rocks of Camajuaní and Remedios Tectonic-Stratigraphic Units (TSU) [UTE]. In south, related to the Regional Belt of Gravimetrical Minima, which coincides with the NCTB [CCNC], a series of intense minimum separated by two zones of maximum of ophiolitic nature are observed. Within the same, it has found maximum Motembo, which, judging by the result of the Ascendant Analytical Continuation to 6 km; the thick of that ultrabasic massif may exceed this magnitude. South and overlapping with the NCTB [CCNC] is Ophiolitic Belt (OB) [CO], characterized by positive values of the field. Within this area stands some minimum related to synorogenic structural depressions (SD) [DE]. Finally, in the southwestern corner of the area an intense maximum linked with a great thickness of volcanic rocks belonging to the Cretaceous Volcanic Arc (CVA) [AVC] is observed.

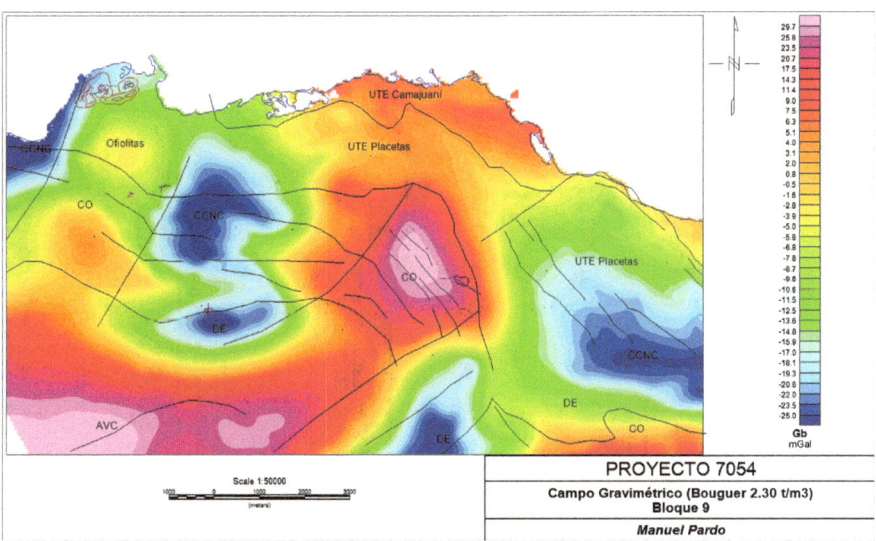

Fig. 3.1 Gravimetric map (Bouguer 2.3 t/m³) at scale 1:50,000 of the Block 9 region

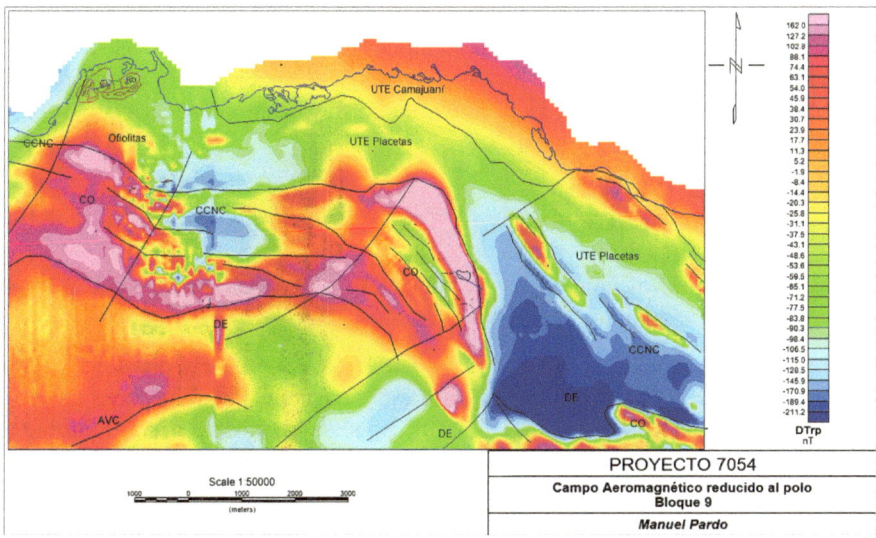

Fig. 3.2 Reduced to pole aeromagnetic map at scale 1:50,000 of the Block 9 region

In the RTP aeromagnetic field shown Fig. 3.2, maximum values are characterized essentially by the ophiolites. For minimum and decreased values, the NCTB [CCNC] and different structural depressions are responsible.

Quantitative estimates from the derivative of the magnetic field inclination (Fairhead et al. 2009) give values of 6.0–7.5 km to the metamorphic basement

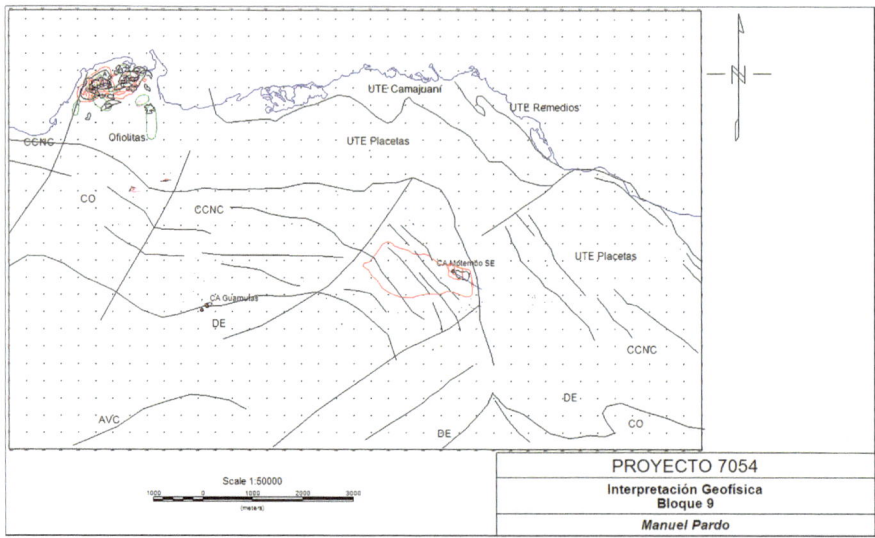

Fig. 3.3 Map of geophysical interpretation of the Block 9 region

below the Remedios TSU [UTE] in the NW area, and 2.0–2.5 km to magnetic targets of CVA [AVC] in the SW and south center of the area.

Thus, at the regional level, the tectonic-structural decipherment of the study area allowed separate the following elements from north to south (Fig. 3.3):

- The Remedios TSU [UTE]
- The Camajuaní TSU [UTE]
- The Placetas TSU [UTE]
- OB and NCTB [CCNC]
- The Cretaceous volcanic rocks (CVR) [RVC] covered by a thick section (2–3 km) of sedimentary rocks, south area.

At the local level, related to the establishment of areas with new possible oil–gas deposits (from the presence of an indicator AC), it should mention three areas of interest: Majaguillar Placetas (TSU [UTE]), Southeastern Motembo (Ophiolites), and Guamutas (northern boundary of Mercedes Basin).

In the area of Majaguillar, the establishment of sectors with oil–gas interest linked to conventional oil of Placetas TSU [UTE], from the presence of a complex of indicator anomalies, considers the following attributes:

- subtle local gravity maximum (green line) (in proximity of regional minimum);
- local magnetic (RTP) highs of very low amplitude (brown line);
- K/Th ratio minimum (red line) and local maxima of U(Ra) on its periphery (pink line); and
- residual relief maximum (black line).

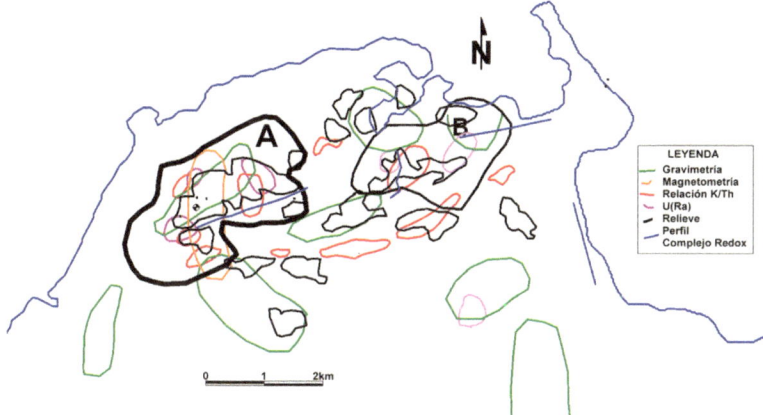

Fig. 3.4 Results of the complex interpretation of geophysical–morphometric methods (non-seismic) in the region of Majaguillar (In *blue line*, recognition profiles of **Redox Complex**)

All these attributes are represented in the complex interpretation results of geophysical–morphometric (non-seismic) methods of Fig. 3.4. Further geological information as the presence of oil wells (black dots) in the region was also considered. According to this figure, the largest oil–gas sector of interest corresponds to an area A (6–7 km^2), northwest region of Majaguillar Swamp, outlined by a regional minimum of the K/Th ratio, and a regional maximum of U(Ra). This sector has the highest number of light hydrocarbon micro seepages in which the largest and concurrent anomalies considered by the different attributes are enclosed. In addition, another area B (4–5 km^2), northeast of this region, is considered, outlined by a regional maximum of U(Ra) and in which are enclosed minor and less concurrent anomalies by the various attributes. This sector has smaller number of light hydrocarbon micro seepages and, therefore, it is considered with a less oil–gas prospectivity. Other sectors (lower area) of possible oil–gas interest are considered those of concurrency of at least two types of attributes, taking, preferably, the subtle local gravimetric maxima as one.

The perspectival sectors for the conventional oil of Placetas TSU [UTE], A and B, have been recognized in land for three profiles of **Redox Complex** (Pardo Echarte and Rodríguez Morán 2016) (one in sector A and two in B; blue lines in Fig. 3.4). There has also been a fourth Redox profile, outside the anomalous area, parallel to the east coast of the region, on a rising of the top of Peñón formation, by seismic data (prospective for unconventional oil of Peñón Fm.) in the locality of the proposed Peñón NC 102 well.

The results of the **Redox Complex** given mainly by local increments of the contents of V, Pb, and Zn validated positively both anomalous results of the complex interpretation of Fig. 3.4 (conventional oil of Placetas TSU [UTE]), as also no anomalous results (unconventional oil of Peñón Fm.) on said the lifting of the top of this last formation.

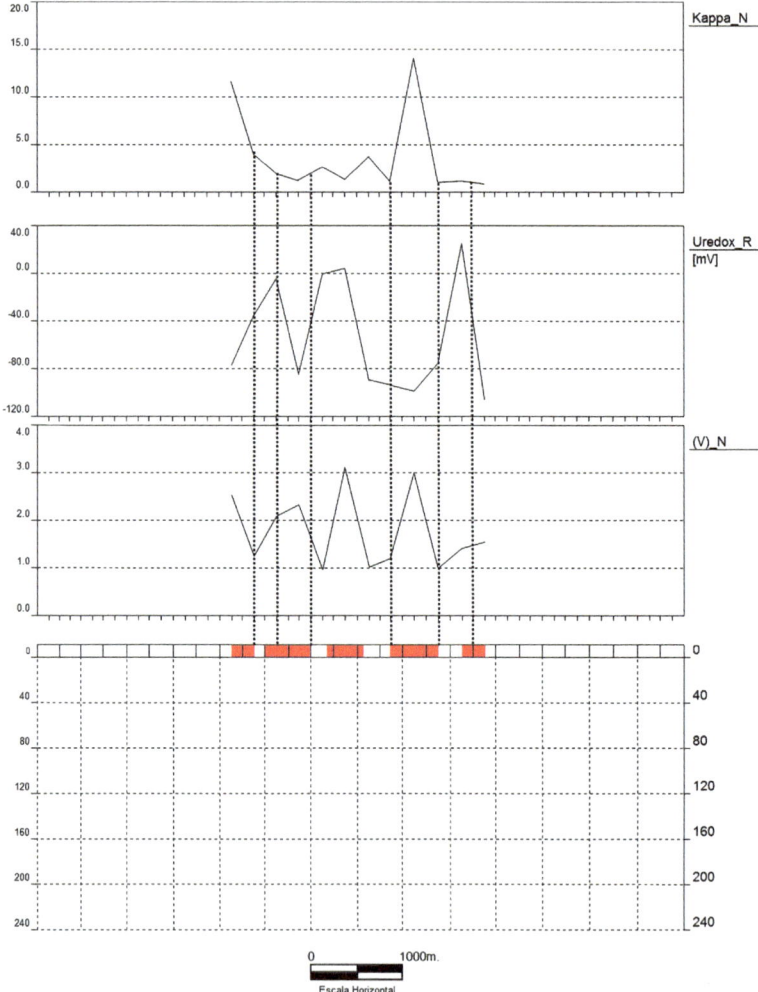

Fig. 3.5 *Redox Complex* profile on the area A, outlook for conventional oil of Placetas TSU [UTE]

In area A, (profile Mj-1 of Fig. 3.5) the profile is anomalous (increments of the contents of V, Pb and Zn) in its entirety (>2000 m), with some typical internal interrupts typically for the processes of light hydrocarbon microseepage in this type of conventional petroleum target.

In the area B, the western profile (profile Majaguillar-ME Fig. 3.6) has a central anomaly (300–400 m long) of V, Pb, and Zn, correlated with increases of magnetic susceptibility and redox potential (possible gas escape) confirming the prospectivity for conventional oil of Placetas TSU [UTE] in that part of the area. In the second profile, further east, on an elevation of the top of the Peñón Fm. (by seismic data)

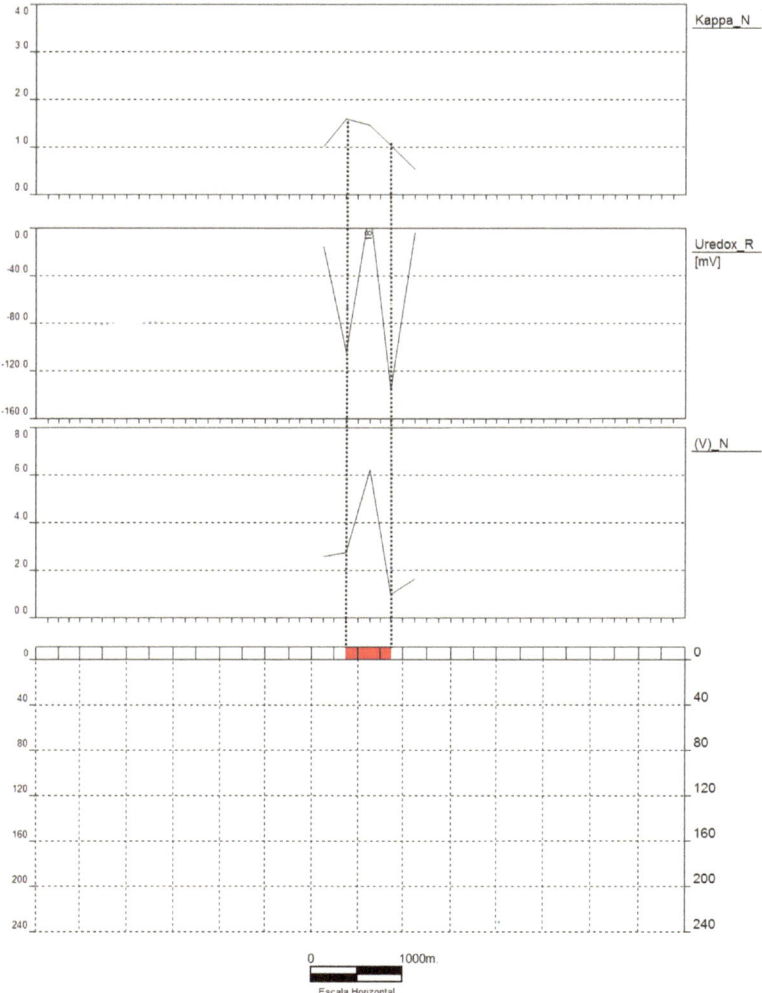

Fig. 3.6 *Redox Complex* western profile over the area B, outlook for conventional oil of Placetas TSU [UTE]

(NC profile 100 of Fig. 3.7), it has a central anomaly (700–800 m long) of V, Pb, and Zn, correlated with increases in magnetic susceptibility and the redox potential (possible gas leak) at its eastern end. This confirms also the prospectivity for conventional oil of Placetas TSU [UTE] and for unconventional oil of Peñón Fm. in that part of the area B.

In the profile parallel to the east coast of the region, on an elevation of the top of Peñón Fm. (by seismic data) (NC profile 102 of Fig. 3.8), a central anomaly (900–1100 m extension) of V, Pb, and Zn, the most intense of the entire region, is also observed. It also correlated with increases of magnetic susceptibility and redox

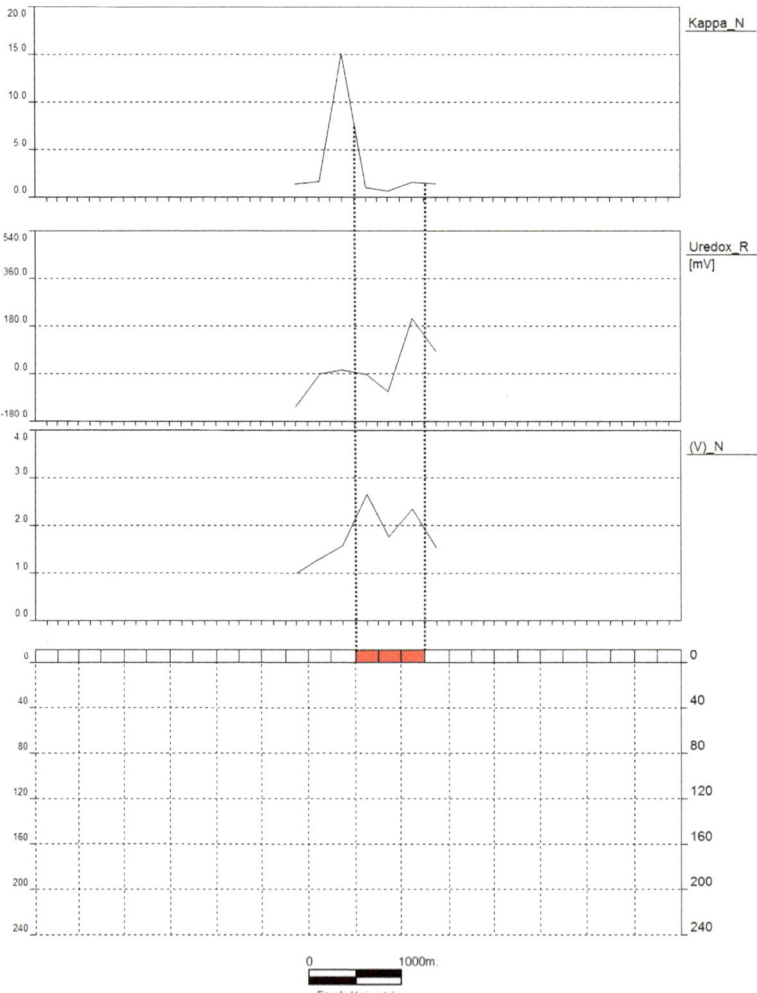

Fig. 3.7 *Redox Complex* oriental profile in the area B, outlook for conventional oil of Placetas TSU [UTE] and for unconventional oil of Peñón Fm., on a lifting of the top of the Peñón Fm., by seismic data

potential (ORP) (possible gas leak). This confirms the perspective for unconventional oil of Peñón Fm. in that part of the area, which result no anomalous by the methods complex used in this study.

Southeast Motembo sector is limited to the east, by a major tectonic dislocation of course N–NW which, apparently, could serve as a migration path for hydrocarbons in the Motembo (*) reservoir area. In this locality, the following AC (Fig. 3.9) is observed:

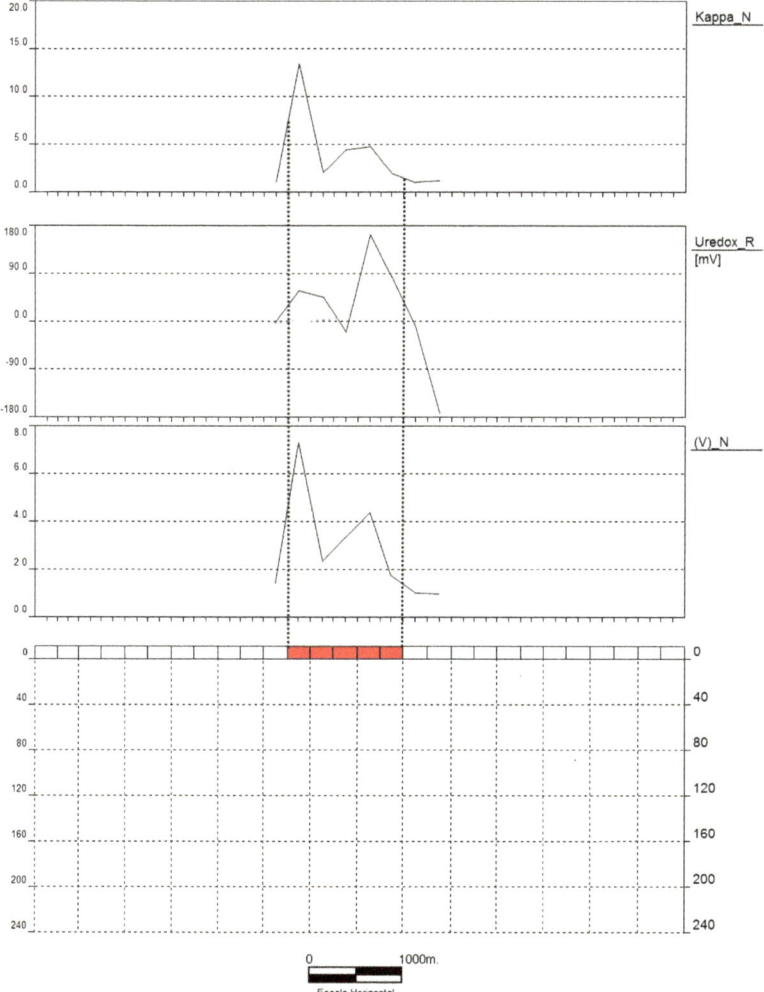

Fig. 3.8 *Redox Complex* profile parallel to the east coast, on an uprising in the top of the Peñón Fm., by seismic data, prospective for unconventional oil of Peñón Fm

- A local gravimetric minimum (green line), oval, coincident with an anomaly (maximum) of residual relief in the order of 10–13 m, in the same manner and position. Judging by its expression in the transformed physical field (Vertical Derivative, Residual Field at 500 and 2000 m) its nature is very superficial.
- A minimum of K/Th ratio (red line) on the western edge of the gravimetric anomaly, least out of it.
- An anomaly of U(Ra) (pink line) on the southern boundary of the K/Th ratio minimum, within it.

Fig. 3.9 Southeast Motembo
Anomalous Complex
(AC) [CA], in the Middle East
Block 9: in *red*, anomaly of
K/Th ratio; in *pink*, maximum
of U(Ra); in *green*, local
minimum Gb-residual relief
maximum; in *black*, tectonic
dislocations; in *blue*, **Redox
Complex** profile; *black dots*
correspond to oil wells

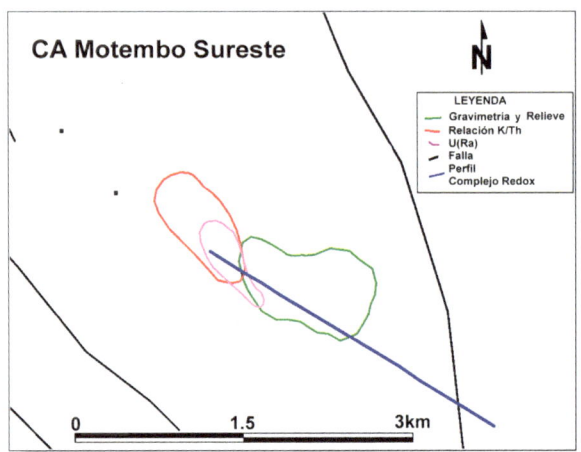

Notably, as interest, the spatial position of this AC is at the very edge of the
ophiolitic massif, where perhaps its thicknesses are not as pronounced, as well as the
uraniferous nature of it, where a possible relationship with hydrocarbons is revealed.

Relatively close to the locality (to the NW) are two oil wells (Vesubio 24
(S/Autor 1954) and Motembo 1X (Sherritt 1995)). The first and next (with a depth
of 381 m) had manifestations of naphtha to 327 m and light oil to 342 m. The
second (with a depth of 1941 m) had slight manifestations of gas. With regard to
the rocks that could constitute seal, the possible target must be found, shallow
(<500 m) in crushed ophiolites, being covered by massive ophiolites or deepest in
rocks of Placetas TSU. The results of the reconnaissance work by the **Redox
Complex** (Fig. 3.10), according to a profile (blue line in Fig. 3.9), were positive to
its western end (on the K/Th and U(Ra) AC) (oil–gas nature expressed by notable
increases of V, Pb, and Zn) and negative about the gravimetric minimum. This
confirms the relationship of the latter with the relief (development of a
ferro-nickeliferous weathering crust). Similarly, a narrow anomalous area (oil–gas
nature, expressed by increments of V, Pb, and Zn) linked to tectonic dislocation
N–NW direction is observed, east of AC, confirming its role as a possible route of
migration of hydrocarbons in the region.

(*) The Motembo site of naphtha (Rodriguez and Kolesnikov 1970; Echevarría
et al. 1991) was discovered in the year 1880–81. The maximum monthly production
was about 26,000 barrels in November 1941. The naphtha accumulation zones
correspond to the serpentine-fractured zones, small size, and with poor or no
communication between them. At present, the deposit is exhausted, so its geophysical–geochemical original expression is not retained (no abnormal expression
is observed).

The Guamutas sector is associated with a major tectonic dislocation sublatitu-
dinal direction (SW–NE) that apparently could serve as a migration path and,
perhaps, turn trap hydrocarbons in this sector. In this locality, the following AC
(Fig. 3.11) is observed:

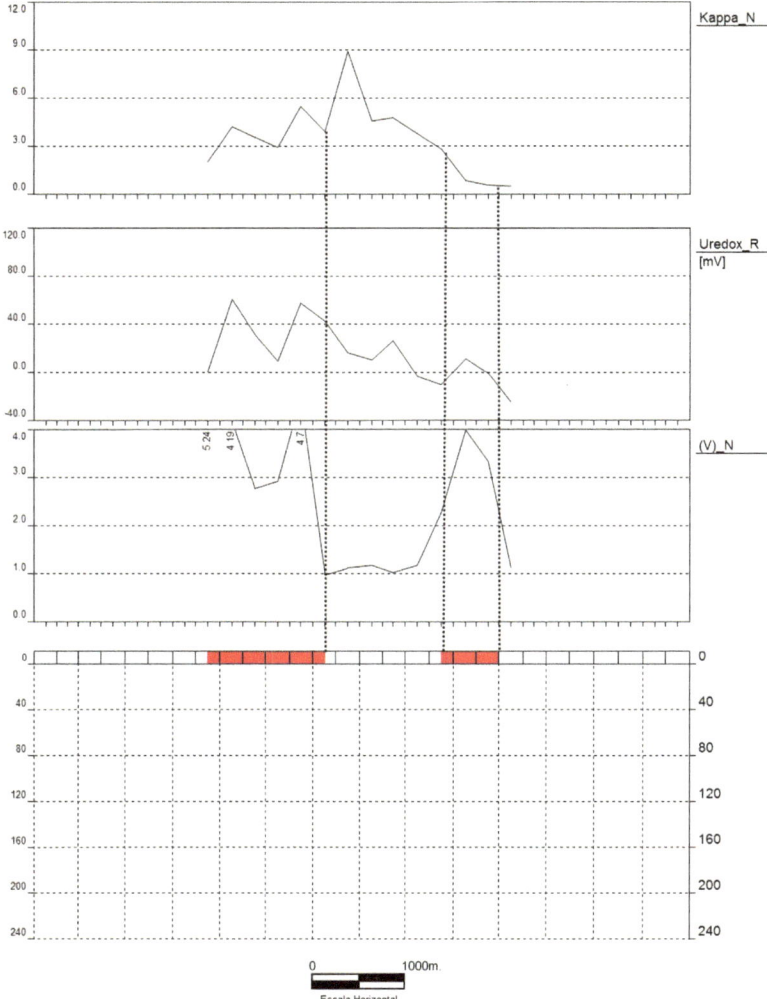

Fig. 3.10 Results of the recognition profile with *Redox Complex* in Southeast Motembo AC [CA] (the distance between points is indicative)

- The northern limit (gradient zone) of a significant Bouguer gravimetric minimum (which reveals an important structural depression—Mercedes Basin, at the South), coincident with an anomaly (maximum) of residual relief in the order of 1.0–1.5 m (green line).
- A local minimum of potassium (red line).
- An area of stressed vegetation SW immediately to the previous AC [CA] (not shown).

Relatively close to the locality (at west), there are four oil wells (West Motembo 1, 2, and 3 (S/Autor 1949) and Peñón 8; of the latter is no information). The closest

Fig. 3.11 Guamutas Anomalous Complex, south center of Block 9: in *red*, minimum of K; in *green*, residual relief maximum; in *black*, tectonic dislocations; in *blue*, **Redox Complex** profile; *black dots* correspond to oil wells

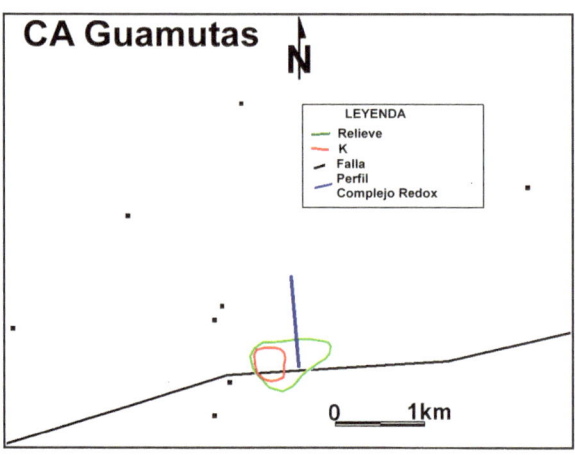

well to the AC [CA] (West Motembo 3), south of tectonic dislocation, with a depth of 634 m, cuts rocks of Oligo-Miocene to 366 m, then rocks of the Eocene–Paleocene to the end of the well, resulting in dry. The well further south, near the area of stressed vegetation (West Motembo 1) 426 m depth, cuts Oligo-Miocene rocks up to 71 m and then serpentine until the end, presenting a manifestation of gas to 327 m. The well at north (West Motembo 2), with a depth of 549 m, cuts Oligo-Miocene rocks up to 295 m, Paleocene–Eocene rocks to 475 m, and then serpentine until the end, resulting in dry. With regard to the rocks that could constitute seal, the possible target must be in the Via Blanca Fm. (Campanian–Maastrichtian), depth <2000 m, being covered by Tertiary rocks with maybe some ophiolite squama and responding to a model of Madruga prospectus of light oil type (Linares et al. 2011). The results of recognition work by the **Redox Complex** (Fig. 3.12), according to a profile (blue line in Fig. 3.11), were positive (oil–gas nature expressed by increases of V, Pb, and Zn; increased magnetic susceptibility and the increase of redox potential, which may be due to a gas leak).

3.3.2   Region of Block 23

The basic geophysical materials for the study of this region are presented in Figs. 3.13 and 3.14 (gravimetric field and RTP aeromagnetics).

In the gravimetric field of Fig. 3.13, highlight the areas of maximum values, the center (Fomentos Maximum), and the southeast area, linked with a powerful thicknesses of CVA [AVC]. By intense minimum values are revealed, in the north-central and eastern area, the NCTB [CCNC] and CB [CC], respectively.

In the RTP aeromagnetic field (Fig. 3.14), mainly the Manicaragua Granitoides (high content of disseminated magnetite), the ophiolites and some varieties of volcanic rocks reveal by maximum values. For minimum values, the Escambray

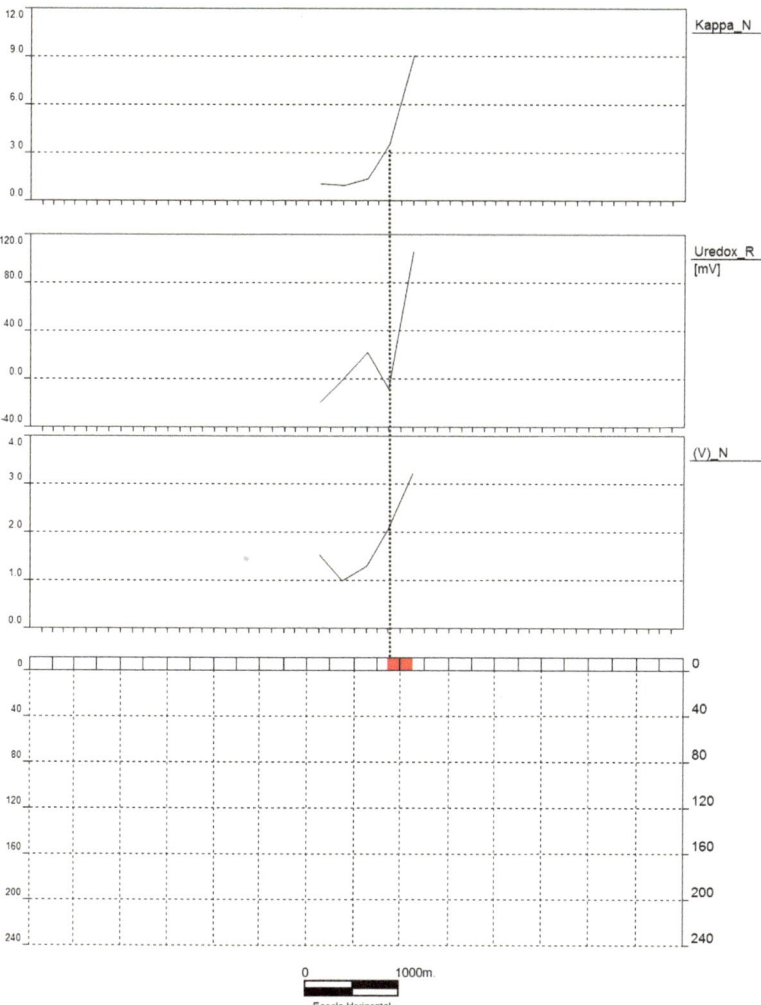

Fig. 3.12 Results of the recognition profile with **Redox Complex** in Guamutas AC [CA] (the distance between points is indicative)

massif metasedimentary rocks and sedimentary rocks corresponding to the NCTB and CB are identified.

At the regional level, the tectonic-structural decipherment of the study region in its central and southeast part allowed to separate the following elements from north to south (Fig. 3.15):

- The Remedios TSU [UTE]
- The Camajuaní TSU [UTE]

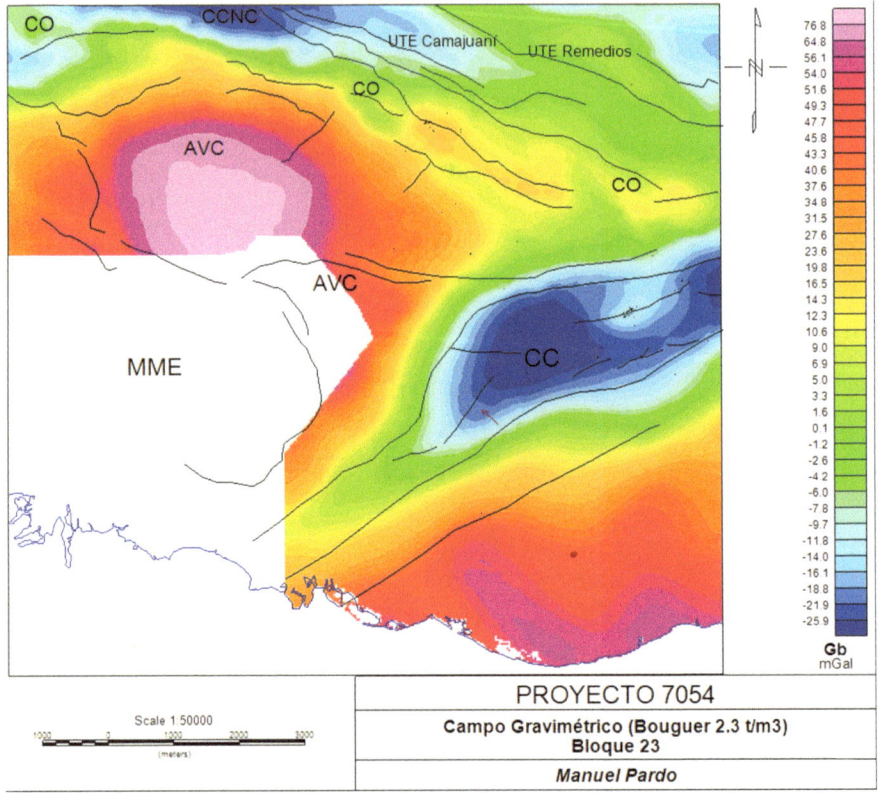

Fig. 3.13 Gravimetric map (Bouguer 2.3 t/m³), scale 1:50 000, of the Block 23 Region

- The OB, which overlaps with the NCTB [CCNC] and the Placetas TSU [UTE] in its southern part
- The volcanic rocks and granitoid belt (GB) [CG] of CVA [AVC]
- The Escambray Metamorphic massif (EMM) [MME]
- The CB [CC].

Quantitative estimates of the depth to magnetic targets derived from the derivative of the total magnetic field inclination (Fairhead et al. 2009) in the CB offer 1500–1800 m values for Jatibonico Maximum and 3200–3600 m to a maximum in the vicinity of the locality of El Pinto (near the Zaza Dam).

At the local level, related to the establishment of areas with new possible oil–gas deposits, it is distinguished in the locality of El Pinto (indicated by the red arrow in Figs. 3.16 and 3.17) with the following AC.

A local circular gravimetric maximum (1200 × 1200 m) with values of −2 mGal of amplitude is spatially coincident with a geomorphic maximum up to 7 m amplitude, with equal dimensions. Judging by its expression in the transformed fields (VD and AAC to 500 and 2000 m), the possible target should be shallow

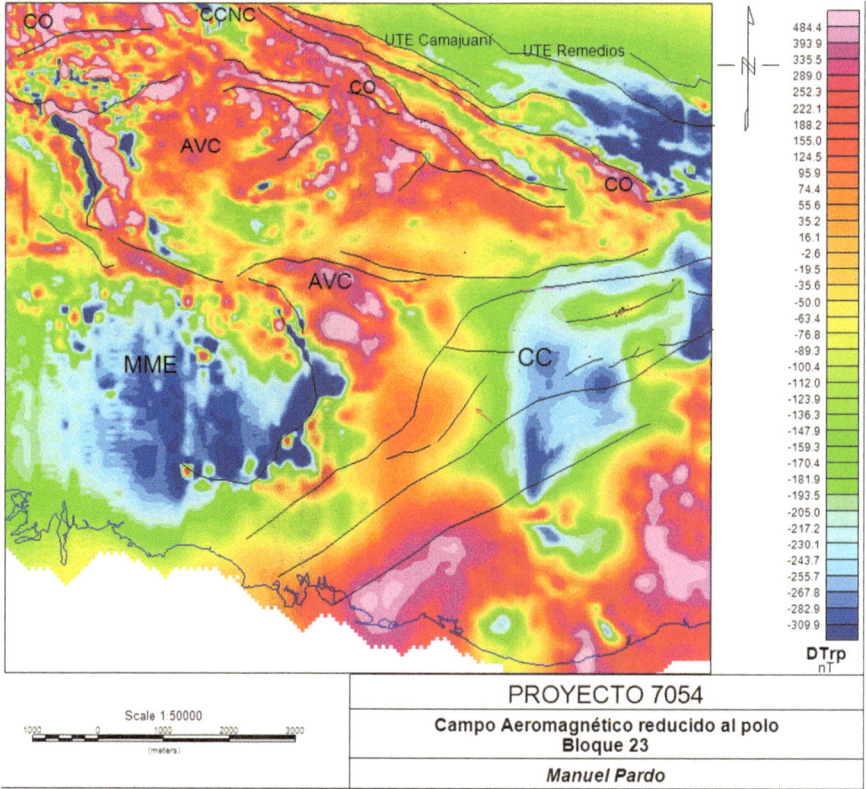

Fig. 3.14 Reduced to pole aeromagnetic map, scale 1:50,000, of the Block 23 Region

(<1000 m), although it is still expressed in the residual gravimetric field at 2000 m. This gravimetrical–geomorphic complex anomaly, on the edge of a tectonic dislocation (similar to Jatibonico Oil Field), has no associated airborne gamma spectrometry indicator response or oil wells; and hydrocarbon shows that are reported next, however, by its nature, presumably linked to hydrocarbon microseepage processes, which must be evaluated by reconnaissance work of *Redox Complex*.

3.3.3   Region of Blocks 17–18

The basic geophysical materials for the study of this region are presented in Figs. 3.18 and 3.19 (gravimetric and RTP aeromagnetic field).

In the gravimetric field of Fig. 3.18, three zones of maximum values are highlighted: Southwest, Central South, and Southeast, linked the first two, with lobes separated· by Amancio Rodríguez Basin, related to powerful thicknesses of

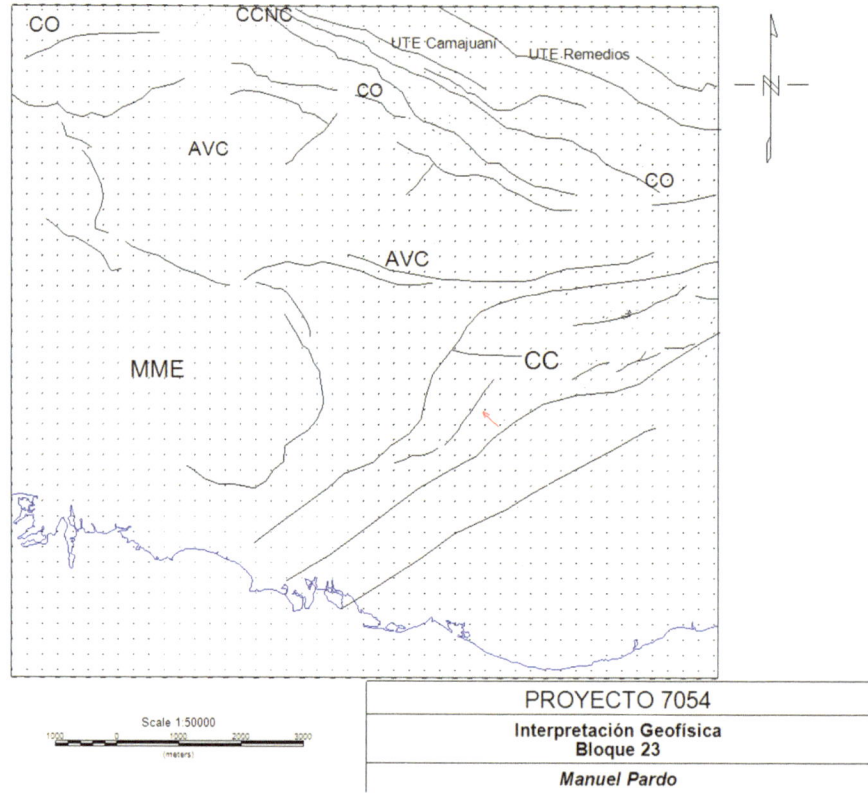

PROYECTO 7054

Interpretación Geofísica
Bloque 23

Manuel Pardo

Scale 1:50000

Fig. 3.15 Geophysical interpretation map, Block 23 Region

volcanogenic-sedimentary rocks of CVA [AVC]. The southeast area, separated from the previous by the Guacanayabo Basin, is ophiolite nature. By intense, minimum values are revealed, to the Northwest, the NCTB and California Basin, on it. Judging by the behavior of the gravimetric field (narrowing of the gravimetric minimum regional belt eastward from Nuevitas Bay), this belt (NCTB [CCNC]) seems wedged south of the region of Gibara.

In the RTP aeromagnetic field of Fig. 3.19, CVA rocks and ophiolites are revealed by maximum values. By minimum values, rocks of Remedios formation and different structural depressions (SD) [DE] are identified.

At the regional level, the tectonic-structural decipherments in the center of the study area allowed separate the following elements from north to south (Fig. 3.20):

- The Remedios TSU [UTE]
- The OB [CO]. This overlaps with the Gravimetric Minimum Regional Belt (linked to the NCTB [CCNC]), which narrower toward the east from the Nuevitas Bay

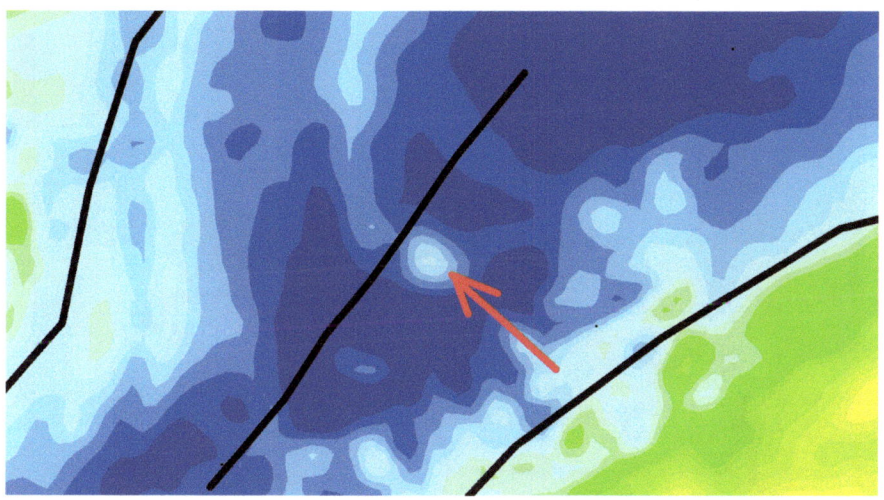

Fig. 3.16 Detail of residual gravimetric field (AAC 500 m) where the local gravimetric maximum is point out in the immediacy of locality El Pinto

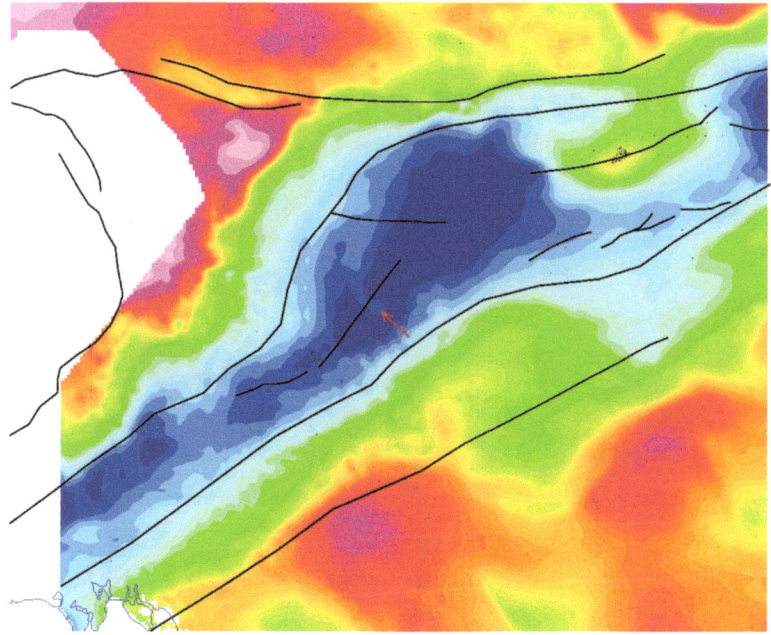

Fig. 3.17 Detail of residual gravimetric field (AAC 2000 m) where the local gravimetric maximum is point out in the immediacy of locality El Pinto

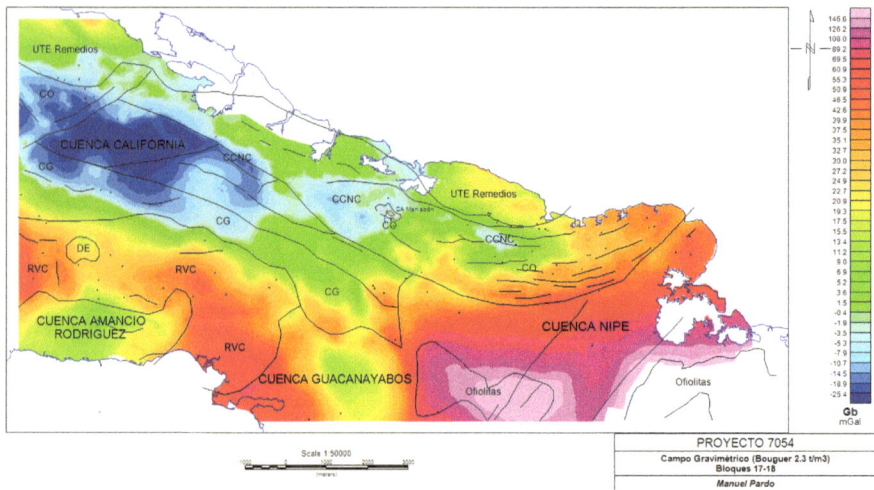

Fig. 3.18 Gravimetric map (Bouguer 2.3 t/m^3), scale 1:50,000, Region of Blocks 17–18

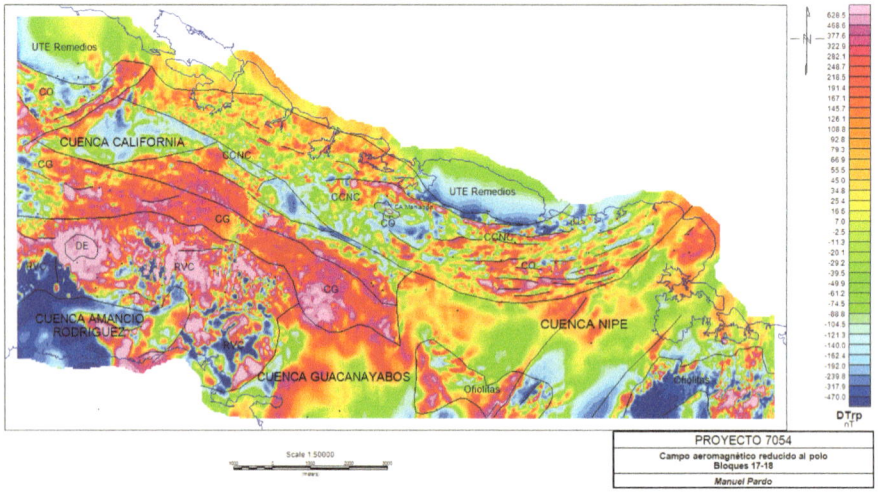

Fig. 3.19 Reduced to pole aeromagnetic map, scale 1:50,000, Region of Blocks 17–18

- The California Basin, on sedimentary rocks from NCTB [CCNC]
- The Granitoids Belt (GB) [CG] of CVA [AVC]
- The CVR [RVC]
- The overlapping depressions on CVR, Amancio Rodriguez and Guacanayabo.

To the west of the study area, separated by the Camagüey Transversal Fault System are distinguished from north to south:

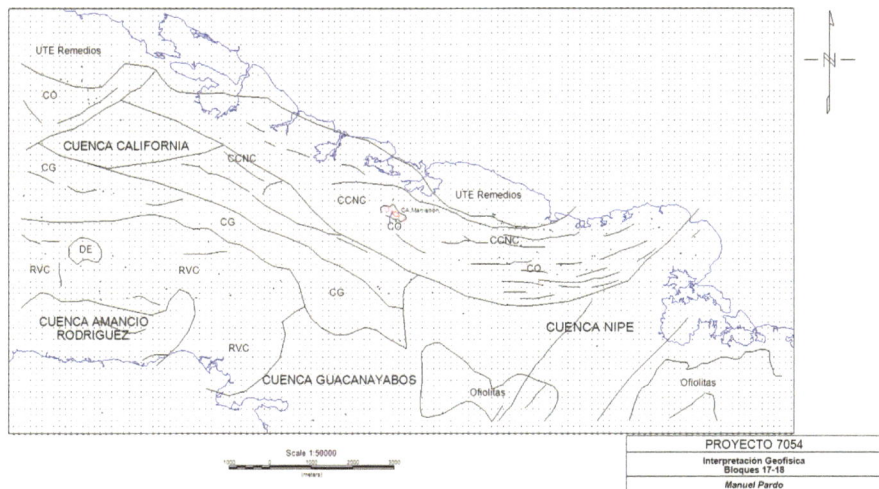

Fig. 3.20 Geophysical interpretation map, scale 1:50,000, Region of Blocks 17–18

- The Remedios TSU [UTE]
- The OB. This overlaps with the Gravimetric Minimum Regional Belt (linked to the NCTB [CCNC]) and Placetas TSU [UTE].

To the east of the study area, south of the OB are distinguished as follows:

- Nipe Basin, on CVR
- Two ultrabasic massifs separated by the Cauto-Nipe Transversal Fault System.

At the local level, related to the establishment of areas with new possible oil–gas deposits, from the presence of a complex of indicator anomalies, it is distinguished at the locality of Maniabón, within foreland basin, by the following AC (Fig. 3.21):

– An oval local gravimetric maximum (red line) with relative values of −1 to −2 mGal within a regional minimum (values between −4 and −5 mGal on Bouguer map). Judging by its expression in the transformed physical fields, the possible target is in the order of 1000 m depth (although the positive structure, apparently inherited from the depth, remains in the residual fields at 2000–6000 m).
– A minimum of K/Th ratio (red line), coinciding with the local gravimetric maximum.
– Three local anomalies of U(Ra) (pink line) in the western and eastern end of AC.
– A positive residual relief anomaly (0.4 m) (red line), coinciding with the local gravimetric maximum.

The Maniabón AC was recognized in ground by works of **Redox Complex** (profile—blue line in Fig. 3.21), made in the area of Templanza–Fortaleza in 2004. It shows a positive anomalous response range of all observed attributes: Kn, 1.5–3;

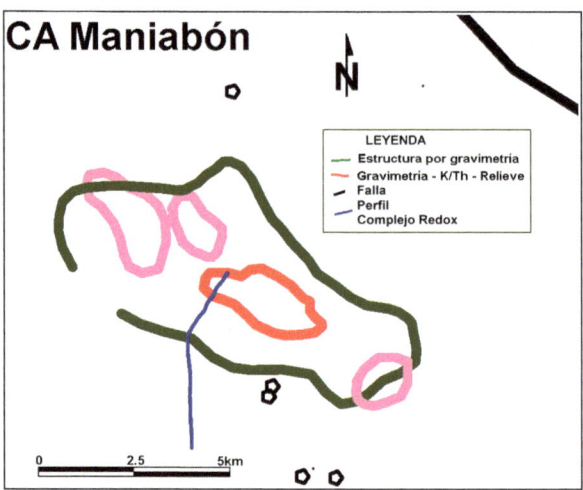

Fig. 3.21 Maniabón anomalous complex, within foreland basin, north center of Blocks 17–18. In *red*, local gravimetric and geomorphic maximums coincident with an anomalous K/Th ratio; in *pink*, maximum of U(Ra); in *green*, outline of the positive structure (by Gb data) circumscribing the local maximum; in *black*, tectonic dislocations; in *blue*, **Redox Complex** profile; *black dots* correspond to oil wells and small pentagons to surface hydrocarbon shows

Ur, less than −30 mV; RER, less than −15%; and Vn (Pb and Zn), 1.5–3; in the share of the local gravimetric maximum (Fig. 3.22).

With regard to the rocks that could constitute seal, the possible target, presumably related with the deposits of the foreland basin, is coated on its western side by Neoauthocton rocks.

In Maniabón sector, numerous reports are about the existence of large surface manifestations of hydrocarbons, which date from 1919, and where heavy oil was extracted from wells with 5 m depth, since colonial times (Linares et al. 2011), it has encouraged oil geologists for exploration of oil and gas deposits in the area.

There have been several exploratory attempts that have coincided in the area of Block17: initiated in 1946 and 1947, the drilling of shallow wells Templanza 1 (495.3 m) and Fortaleza 1 (304.2 m), motivated by the abundant presence of shallow heavy oil manifestations in the Maniabón sector, which produced small volumes of oil and gas. Later there were drilled, north of the area of the manifestations, the well Puerto Padre 1, in 1958 (1099.2 m), and Farola Norte 1, in 1998 (2314 m), without productive results; up to the recent drilling, in 2011 of Picanes 1X well (3568 m), toward the center of the block. The latter, in the final stages of drilling, showed considerable gas inlets into the mud, reaching measure in the order of 60% of total gas (Pérez et al. 2013). The aforementioned aspects are irrefutable evidence of the existence of elements of an active petroleum system in the whole area.

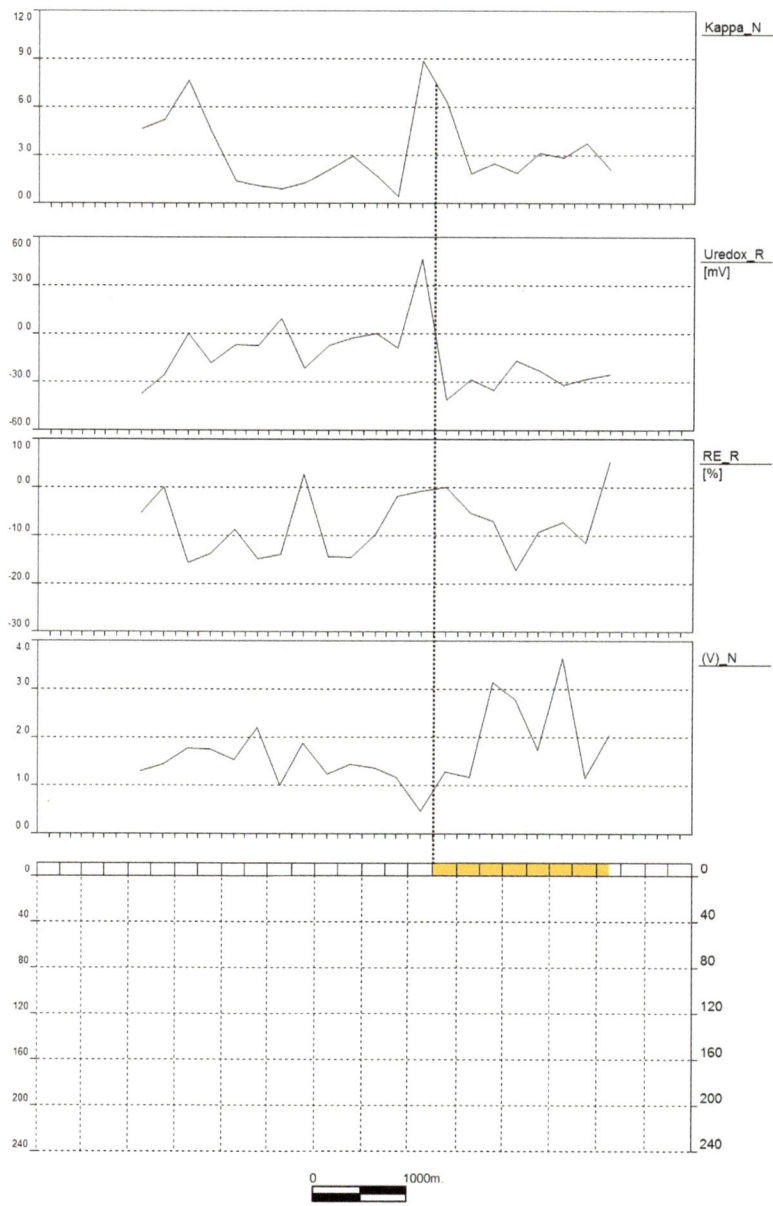

Fig. 3.22 Results of *Redox Complex* reconnaissance profile in the area of Templanza-Fortaleza, partially on Maniabón anomalous complex

3.4 Conclusions

A version of the tectonic-structural regionalization of different territories under the study based on potential fields, having as main references the VD map of the magnetic and gravimetric fields and together with an analysis of the remaining transformations of the fields, is offered. The chapter presents a summary of the main results achieved in the tectonic-structural regionalization with purposes of hydrocarbon exploration, focusing on the mapping of new possible oil–gas objectives in the regions of Onshore Blocks 9, 23, and 17–18. At the local level, attention is paid to AC indicators of new possible deposits:

– local gravimetric and geomorphic highs within or in the vicinity of regional gravimetric minimum;
– minimum of K/Th ratio and local anomalies of U(Ra) in the periphery;

that are located in Northern Cuba, within the overlapping area of Gravimetric Minimum Regional Belt (linked to the NCTB [CCNC]) with OB, lying on the Placetas TSU [UTE]. In the southwestern region of CB, attention is paid to the local gravimetric and geomorphic maximums (similar to Jatibonico Oil Field).

In some of the new places of interest (Majaguillar, Southeast Motembo, Guamutas, and Maniabón), reconnaissance works were carried out by a profile of **Redox Complex** (complex of unconventional geophysical–geochemical exploration techniques) with positive results. The new areas proposed should be evaluated from the seismic point of view to clarify the structural aspects that allow the location of possible exploration wells.

In Block 9, from the presence of a complex of indicator anomalies, three locations of greater oil–gas interest were established: Majaguillar (Placetas Tectonic-Stratigraphic Unit), Southeast Motembo (Ophiolites), and Guamutas (northern boundary of Mercedes Basin). Block 23, in the locality of El Pinto (near the Zaza Dam), is distinguished as an AC given by a local gravimetric maximum, coincident with a geomorphic maximum with the same position and dimensions. Blocks 17 - 18 are distinguished, in the locality of Maniabón (south of Puerto Padre and Nuevitas bays), as an AC given by a local gravimetric–geomorphic maximum within a regional gravimetric minimum, a minimum of K/Th ratio, and local U(Ra) anomalies at the ends of the AC.

Definitively, for its geological significance (associated to foreland basin), it was established as main interest of all locations, the Maniabón area.

References

Colectivo de Autores (2007) Mapa Geológico Digital con fines petroleros de la República de Cuba a escala 1:250,000. Inédito. Centro de Investigaciones del Petróleo, La Habana
Colectivo de Autores (2008) Mapa Digital de las Manifestaciones de Hidrocarburos de la República de Cuba a escala 1:250,000. Inédito. Centro de Investigaciones del Petróleo, La Habana

Colectivo de Autores (2009) Mapa Digital de los Pozos Petroleros de la República de Cuba a escala 1:250,000. Inédito. Centro de Investigaciones del Petróleo, La Habana

Colectivo de Autores (2010) Mapa Geológico Digital de la República de Cuba a escala 1:100,000. Inédito. Instituto de Geología y Paleontología, Servicio Geológico de Cuba, La Habana

Echevarría G, Hernández-Pérez G, López-Quintero JO, López-Rivera JG, Rodríguez-Hernández R, Sánchez-Arango J, Socorro-Trujillo R, Tenreyro-Pérez R, Iparraguirre-Pena JL (1991) Oil and gas exploration in Cuba. J Pet Geol 14(3):259–274

Fairhead JD, Ahmed Salem and Williams SE (2009) Tilt-depth: a simple depth-estimation method using first order magnetic derivatives. Search and Discovery Article#40390 (Adapted from poster presentation at AAPG International Conference and Exhibition, Cape Town, South Africa, October 26–29, 2008)

Linares E, García Delgado D, Delgado López O, López-Rivera JG y Strazhevich V (2011) Yacimientos y manifestaciones de hidrocarburos de la República de Cuba. Centro de Investigaciones del Petróleo, La Habana, 480 págs

Mondelo F Sánchez R y otros (2011) Mapas geofísicos regionales de gravimetría, magnetometría, intensidad y espectrometría gamma de la República de Cuba, escalas 1: 2,000,000 hasta 1: 50,000. Inédito. IGP, La Habana, 278p

Pérez-Martínez Y y otros (2013) Proyecto 7054. Etapa 1.4. Informe final sobre fundamentación de pozo en el Bloque 13. Archivo técnico CEINPET

Pardo Echarte ME and Rodríguez Morán (2016) Unconventional methods for oil & gas exploration in Cuba. Springer Briefs in Earth System Sciences. doi10.1007/978-3-319-28017-2

Rodríguez R y Kolesnikov L (1970) Informe sobre el área de Motembo y Corralillo. Inédito. Archivo CEINPET O-45, 53 p

S/Autor (1949) Columnas lito paleontológicas de los pozos West Motembo 1, 2 y 3. Inédito. Archivo CEINPET 91955, 3 p

S/Autor (1954) Columnas lito paleontológicas de los pozos Vesubio 24, 25 y 26. Inédito. Archivo CEINPET 91950, 3 p

Sherritt (1995) Informe sobre el pozo Motembo 1X. Inédito. Archivo CEINPET E-260, 7 p

Verduzco B, Fairhead JD, Green CM, MacKenzie C (2004) New insights into magnetic derivatives for structural mapping. Lead Edge 23(2):116–119

https://lpdaac.usgs.gov/ Modelo Digital de Elevación 30 × 30 m de la República de Cuba

Chapter 4
Geological-Structural Mapping of Weathered Igneous and Metamorphic Rock Units

Abstract A summary of the main results achieved in the geological-structural mapping, from the data of potential fields and airborne gamma spectrometry, from units of igneous and metamorphic rocks in the western (Havana-Matanzas), central (Cienfuegos-Villa Clara-Sancti Spiritus), and central-eastern (Camagüey-Tunas-Holguín) regions of Cuba offers. The gravimetric data allow to identify different geological-structural traits: by lows, those associated with the Northern Cuban Thrusts Belt, the southern metamorphic massifs, the granitic igneous bodies and the synorogenic structural basins and depressions; by highs, those linked with powerful thickness of volcanic rocks and ophiolitic bodies; as well as geophysical alignments, major tectonic boundaries within the Cuban Orogen. Aeromagnetic data allow mapping the main tectonic boundaries; the southern metamorphic massifs; the synorogenic structural basins and depressions; granitoid belts; ophiolitic bodies and the development area of volcanic rocks. The faculty of lithological mapping that gives the differential distribution of magnetite in various rock units gives this possibility. The airborne gamma spectrometry identifies, by increased values of U(Ra), units with high contents of graphite (organic matter) and those associated with acid igneous rocks. Potassium increases are linked, mainly, to the alkaline and acidic medium-igneous rocks. Incremented values of thorium generally characterize the metamorphites and, U(Ra) and Th increments express some highly developed weathering crusts on ultrabasites.

Keywords Geological-structural mapping · Lithological mapping · Gravimetry · Aeromagnetics · Airborne gamma spectrometry · Igneous and metamorphic rocks

4.1 Introduction

Cuban geological history and tropical climate determine the presence of contrasting residual soils or in situ weathering crusts and an eminently plain relief. Cuba has, in addition, an aeromagnetic and airborne gamma spectrometry survey at scale 1:50,000 from throughout the country and a gravimetric survey, 80% of it on the

© The Author(s) 2017
M.E. Pardo Echarte and J.L. Cobiella Reguera, *Oil and Gas Exploration in Cuba*,
SpringerBriefs in Earth System Sciences, DOI 10.1007/978-3-319-56744-0_4

same scale. Within the regional geophysical investigations that are medium to large scale (1:250,000–1:50,000) undertaken in Cuba at present, are those related to the mapping of weathering mantles corresponding to units of igneous and metamorphic rocks, for the Mineragenic Map of Cuba scale 1:250,000. The geological task posed to processing and interpreting geological–geophysical data consisted of tectonic-structural decipherment of the investigated regions (Havana-Matanzas, Cienfuegos-Villa Clara-Sancti Spiritus and Camagüey-Las Tunas-Holguín) from potential fields and, mapping of weathered igneous and metamorphic rock units from the airborne gamma spectrometry data. To this end, the gravimetric, aeromagnetic, and airborne gamma spectrometry fields of the different territories are processed. Whenever aspects of Cuban geology are conveniently discussed in chapter of Geological Overview, the results for each study region are limited to present basic materials and their regional geophysical interpretation with a brief description.

4.2 Materials and Methods

4.2.1 Materials

In carrying out the investigation, the following were used as primary sources of information:

- Grids of gravimetric and aeromagnetic fields to 1:50,000 scale and, airborne gamma spectrometry (channels: It, U, Th and K) at 1:100,000 scale of the Republic of Cuba (Mondelo et al. 2011).
- Digital Geological Map of the Republic of Cuba at scale 1:100,000 (Colectivo de Autores 2010).

4.2.2 Methods and Techniques

Methods

It is an theoretical–empirical research, using the hypothetical-deductive method, where, from the numerical information of physical fields and through operations of analysis and deduction, the complex phenomenon of geology and geophysics of the study regions decomposes in their integrant parts, structure-tectonic and lithological, in order to define its essential features under the stated hypotheses.

Techniques

The Automated Processing and Interpretation System of Geophysical-Geological Data, Oasis Montaj (version 7.01) was used: automated processing of geophysical–

geological georeferenced information (matrix arithmetic operations, transformations of physical fields and others) are performed.

Processing and Interpretation

The gravimetric field, for the purposes of geological-structural mapping of weathered metamorphic and igneous rock units, was submitted to the regional-residual separation from the Ascendant Analytical Continuation (AAC) for the height of 5000 m, given by the order of average depth of the deep structure in the study regions. Vertical Derivative (VD) was also calculated for the establishment of local anomalies that would allow a more accurate mapping of the territory tectonic- structural limits.

The magnetic field was submitted to the VD and the total horizontal derivative of the tilt derivative of the field (Verduzco et al. 2004), decisive for tectonic-structural decipherment of the territory. In addition, the reduction to pole and the regional field analysis (AAC for the height of 5000 m) was made. It highlights that the tectonic limits fixed by the chains axes of highs from total horizontal derivative of the tilt derivative of the field match, in general, with the chains axes of minimums from the vertical derivative of the field (reduced to pole). Only, the latter offers a much less noisy structural frame and, therefore, a more detailed and preferred version.

The airborne gamma spectrometry considered the analysis of each channel, relations between channels (K/Th, Th/K, K/U, U/K, U/Th, and Th/U) and ternary maps plus the combination of these with different relations.

Geophysical alignments, regional tectonic dislocations that separate different tectonic-structural elements were drawn, mainly, from the VD maps of the magnetic and gravimetric fields; considering chains of minimums (mainly) and maximums, linearity, push-ups, and interruption of contour lines and areas of high gradient of them.

The position of the alignments and limits outlined from geophysical maps of the VD of potential fields is considered acceptable in comparison with the corresponding geological map. The character depth of these limits is established, in general, as vertical–subvertical (which could be softer with depth); generally dip to the south, to the north of Cuba, from comparing its position regarding the observed and processed fields of different nature.

Similarly, the limits of the main depressions considering a change in the characteristics of the VD of magnetic field (smoothing or flattening), coincident with regional gravitational minimum were plotted.

The mapping of different weathered metamorphic and magmatic lithological units was performed, mainly, from the different ternary maps drawn and considered the areas of minimum and maximum values of integral channel.

4.3   Results

The contribution of geophysical data to the geological-structural mapping of the different study regions in Cuba meets the well-established regularity that potential fields help, basically, to the tectonic-structural decipherment of the territory and, to a lesser extent, to lithological mapping of the different units present; resulting in reverse the contribution of airborne gamma spectrometry data.

The gravimetric data allow to identify different geological-structural traits: by lows, those associated with the Northern Cuban Thrusts Belt, the southern metamorphic massifs, the granitic igneous bodies and the synorogenic structural basins and depressions; by highs, those linked with powerful thickness of volcanic rocks and ophiolitic bodies; as well as geophysical alignments, major tectonic boundaries within the Cuban Orogen. Aeromagnetic data allow mapping the main tectonic boundaries; the southern metamorphic massifs; the synorogenic structural basins and depressions; granitoid belts; ophiolitic bodies and the development area of volcanic rocks. The faculty of lithological mapping that gives the differential distribution of magnetite in various rock units gives this possibility. The airborne gamma spectrometry identifies, by increased values of U(Ra), units with high contents of graphite (organic matter) and those associated with acid igneous rocks. Potassium increases are linked, mainly, to the alkaline and acidic medium-igneous rocks. Incremented values of thorium generally characterize the metamorphites and, U(Ra) and Th increments express some highly developed weathering crusts on ultrabasites.

4.3.1   Central Region

The basic geophysical materials for the study of this region are presented in Figs. 4.1, 4.2 and 4.3 (gravimetric field, reduced to pole aeromagnetics and Ternary Map).

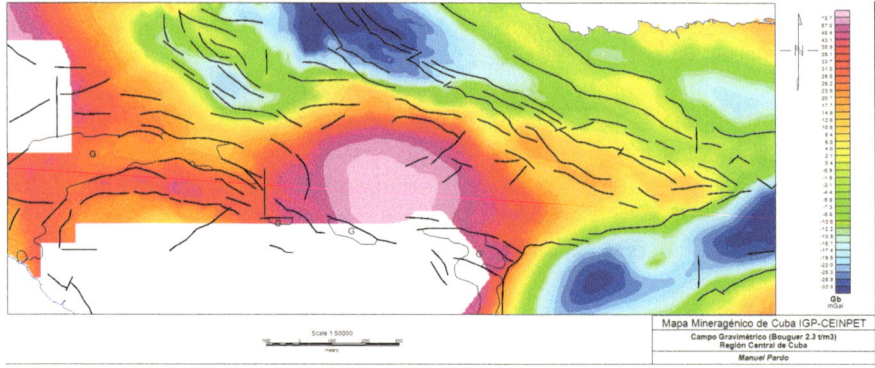

Fig. 4.1 Gravimetric map (Bouguer 2.3 t/m^3), scale 1:50,000, central region

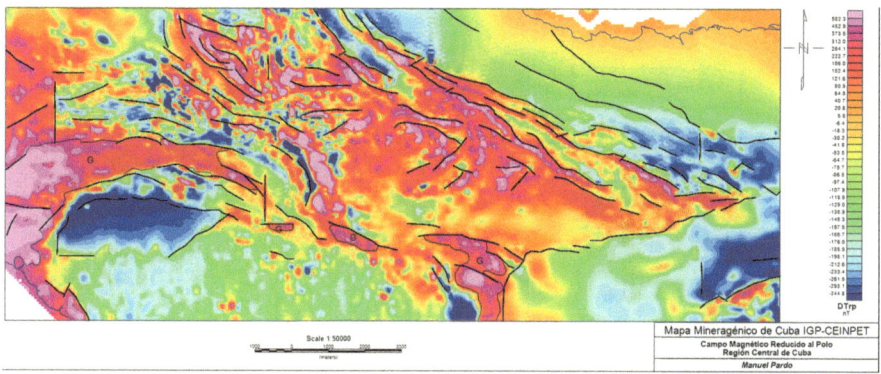

Fig. 4.2 Reduced to pole aeromagnetic map, scale 1:50,000, central region

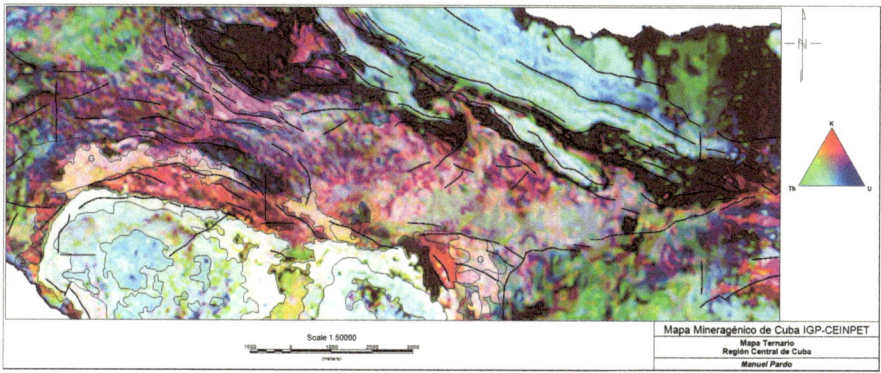

Fig. 4.3 Ternary map, scale 1:100,000, central region

The gravimetric field of Fig. 4.1 reveals two zones of maximum values, west and center of the area (Cienfuegos and Fomentos maximums, respectively), linked with powerful thickness of Cretaceous volcanic rocks. A narrow belt of increased values, north of Escambray Metamorphic Massif, is related to the Mabujina Amphibolitic Complex, which, judging by the results of the Ascendant Analytical Continuation of the field at 5000 m, it is wedged with depth. By intense minimum values, at the North-Central and Southeast area, the Northern Cuban Thrusts Belt and the Central Basin, respectively, are revealed.

In the reduced to pole aeromagnetic field of Fig. 4.2, maximum values are revealed, mainly, by the Manicaragua granitoids (from the high content of disseminated magnetite), the ophiolites and some varieties of volcanic rocks. For minimum values, the Escambray massif metasedimentary rocks and sedimentary rocks from Northern Cuban Thrusts Belt and Central Basin are identified.

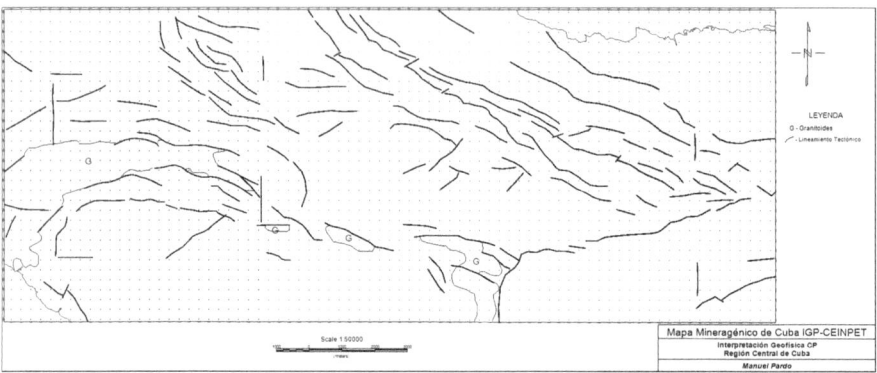

Fig. 4.4 Map of geophysical interpretation, Potential Fields, central region

In the Ternary Map of Fig. 4.3 is identified: in whitish tones, the Escambray Metamorphic Massif. In pinkish tones, Cretaceous volcanic arcs. In dark colors, ophiolitic rocks, the basalts of Los Pasos Formation and amphibolites. In bluish tones, carbonate rocks of Bahamas Continental Margin (Remedios and Camajuaní tectonic-structural units and Escambray Metamorphic Massif (San Juan Fm.) and; in green tones, other sedimentary rocks from Neoauthocton cover.

At the regional level, the tectonic-structural decipherment of the territory shows different geological-structural features and tectonic-structural units (Fig. 4.4):

- East-southeast of the area, La Trocha Fault Zone (Central Basin).

From north to south,

- The Remedios, Camajuaní and Placetas Tectonic-Stratigraphic Units.
- The Ophiolitic Belt.
- The Cretaceous volcanic arcs.
- The Escambray Metamorphic Massif.

From the lithological mapping point of view, from the airborne gamma spectrometry data (Fig. 4.5) can be recognized in different ternary maps, weathering crusts from different geological units:

- Mabujina Amphibolites, Loma La Gloria, Cobrito and San Juan formations, in the Escambray Metamorphic Massif;
- The main granitoid bodies, basalts of Los Pasos Fm. and acid tuffs of Bruja Fm., in the Cretaceous volcanic arcs;
- The ophiolites (bodies of serpentinized ultramafites and gabbros).

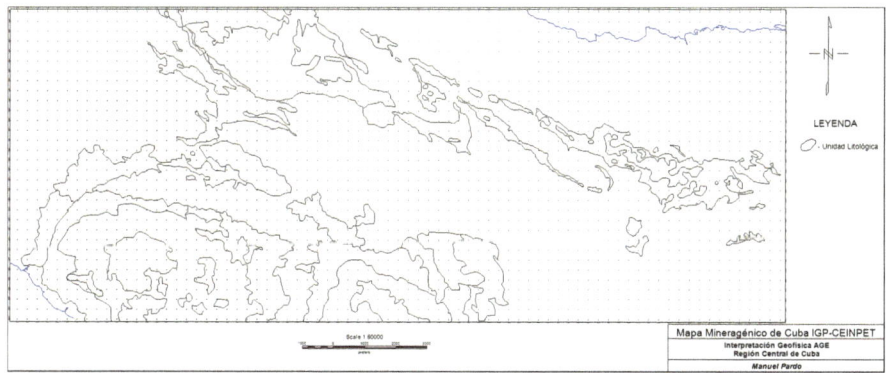

Fig. 4.5 Map of geophysical interpretation, Airborne Gamma Spectrometry, central region

4.3.2   Camagüey-Las Tunas-Holguín Region

The basic geophysical materials for the study of this region are presented in Figs. 4.6, 4.7, and 4.8 (gravimetric field, reduced to pole aeromagnetics and Ternary Map).

The gravimetric field of Fig. 4.6 reveals three zones of maximum values: West, Central South, and Southeast; linked, the first, with the of Cretaceous island arc granitoids belt (Gaspar Maximum), the second (with two lobes separated by Amancio Rodríguez Basin) related to a powerful thicknesses of the same arc volcanic rocks. The Southeast region is ophiolitic nature. By intense minimum values are revealed, at the center-north, the Northern Cuban Thrusts Belt and California Basin on it. Judging by the behavior of the gravimetric field, this belt seems wedged south of the Gibara region. Southwest of Camagüey transverse tectonic corridor, a minimum gravimetric area linked to the Vertientes Basin is distinguished.

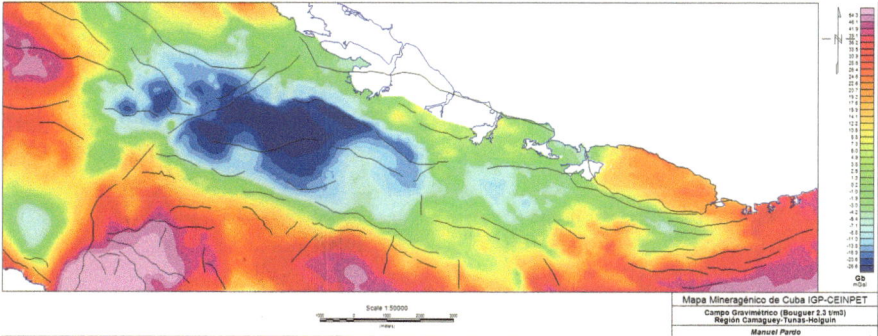

Fig. 4.6 Gravimetric map (Bouguer 2.3 t/m^3), scale 1:50,000, Camagüey-Las Tunas-Holguín region

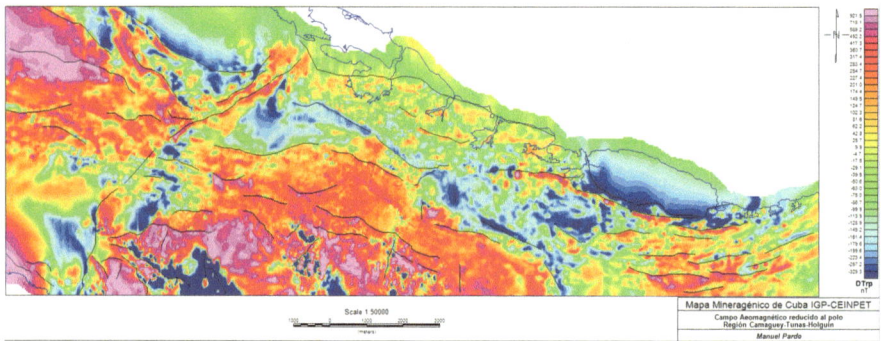

Fig. 4.7 Reduced to pole aeromagnetic map, scale 1:50,000, Camagüey-Las Tunas-Holguín region

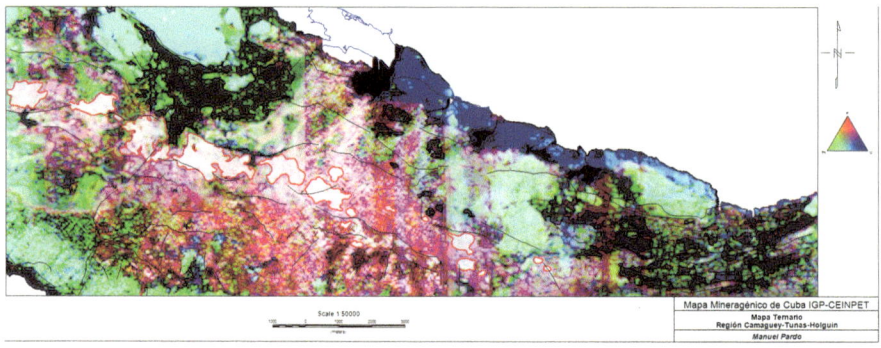

Fig. 4.8 Ternary map, scale 1:100,000, Camagüey-Las Tunas-Holguín region

In the reduced to pole aeromagnetic field of Fig. 4.7, are revealed, by peaks, Cretaceous volcanic arc rocks and ophiolites. For minimum values, rocks of Remedios Fm. and different structural depressions are identified.

In the Ternary map of Fig. 4.8, in pinkish tones, the Cretaceous volcanic arc rocks are revealed and, within them, the whitish tones identify some granitoids bodies, most probably, from the granosienitic and alkaline leucocratic granite complexes. In bluish, clear tones, carbonate rocks of the Remedios Fm. are revealed, and in dark tones, carbonate rocks of the Jaimanitas Fm. With dark-black tones, ophiolites are identified and, in green tones, other sedimentary rocks from Neoauthocton cover.

At the regional level, the tectonic-structural decipherment of the territory shows different structural features and units (Fig. 4.9), from north to south:

Remedios Tectonic-Stratigraphic Unit [A]
Ophiolitic Belt [B]
Granitoids Belt [C]
Volcanogenic Sedimentary Complex of Cretaceous volcanic arc [D]

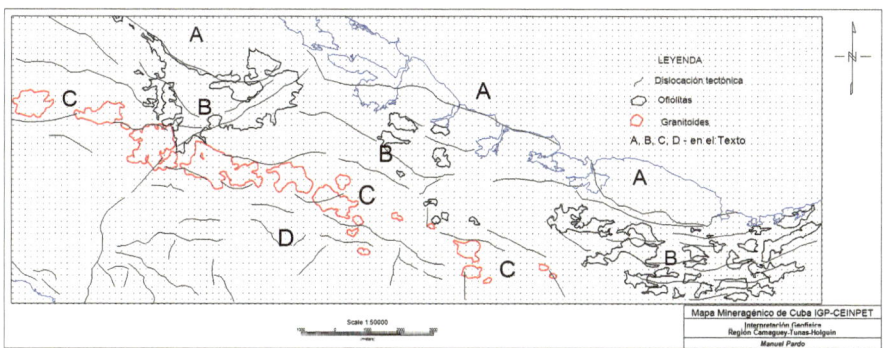

Fig. 4.9 Map of geophysical interpretation, Camagüey-Las Tunas-Holguín region

As a cross-tectonic fault zone, Camagüey corridor stands.

From the lithological-mapping point of view, from the airborne gamma spectrometry data can recognize, only, the weathering crusts of the ophiolitic rocks (sharp decline in the response by the three channels) and certain types of granitoids (simultaneous abnormal response by the three channels). The volcanogenic sedimentary rocks are generally indistinguishable from the airborne gamma spectrometry data; although areas of stronger pink tones could correspond with the alkaline varieties (Camujiro and Piragua formations). Within sedimentary rocks stands, on the north coast, the Fm. Jaimanitas by abnormal response in channel U(Ra).

4.3.3   Habana-Matanzas Region

The basic geophysical materials for the study of this region are presented in Figs. 4.10, 4.11 and 4.12 (gravimetric field, reduced to pole aeromagnetics and Ternary Map).

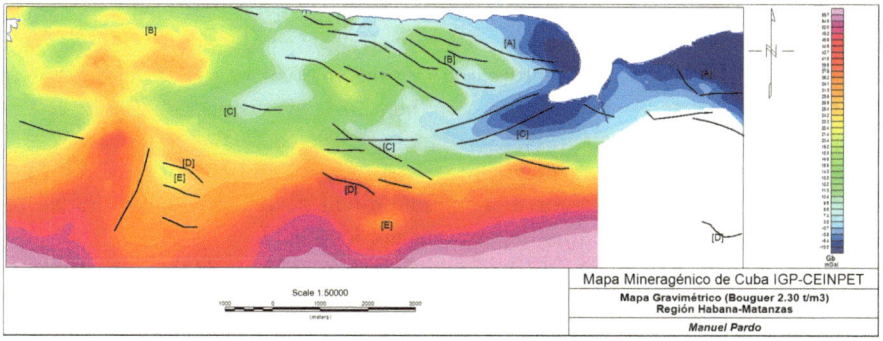

Fig. 4.10 Gravimetric map (Bouguer 2.3 t/m^3), scale 1:50,000, Habana-Matanzas region

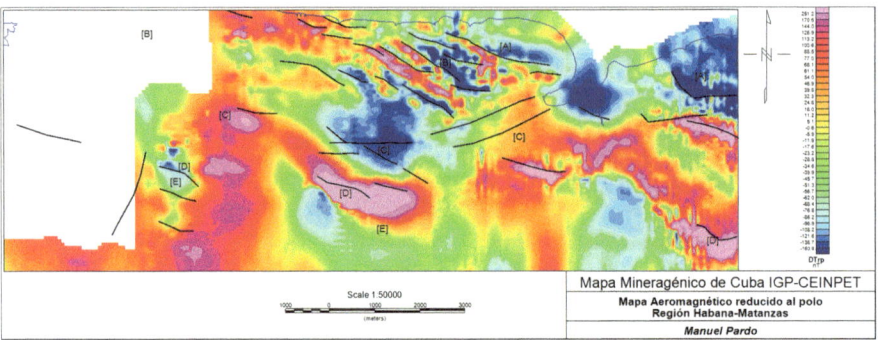

Fig. 4.11 Reduced to pole aeromagnetic map, scale 1:50,000, Habana-Matanzas region

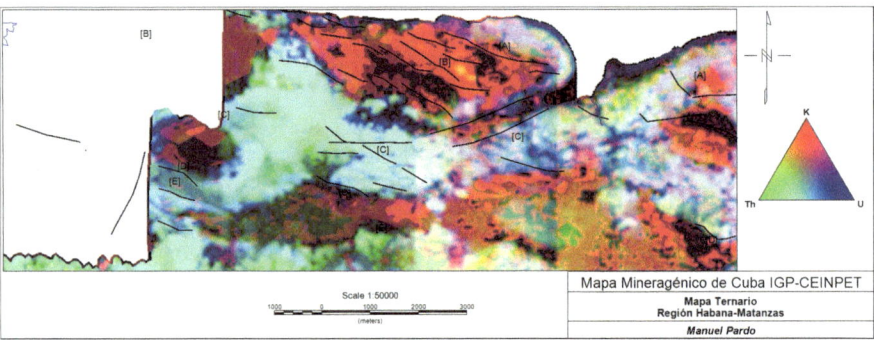

Fig. 4.12 Ternary map, scale 1:100,000, Habana-Matanzas region

In the gravimetric field of Fig. 4.10 is shown, south, an area of maximum values linked with a powerful thickness of volcanogenic sedimentary rocks of Cretaceous volcanic arc. By intense minimum values is revealed, northeast, the Northern Cuban Thrusts Belt (NCTB) [CCNC], with exit to the sea. There are also distinguished by increased intermediate values, south of the NCTB [CCNC], a narrow belt of ophiolites [B] and another, southern [D] (Madruga-Bejucal elevations), separated by a central region of decreased intermediate values [C] corresponding to the Mercedes-Aguacate basin area. Another area of decreased values [E] is observed south of the southern belt of ophiolites, related to Vega Basin.

In the reduced to pole aeromagnetic field of Fig. 4.11, ophiolites are revealed, essentially, by maximum values. For minimum and decreased values, the NCTB [CCNC] and different structural depressions are identified.

In the Ternary Map of Fig. 4.12, in dark-black and reddish tones, the ophiolites of the northern belt mixed with Chirino volcanics formation of Cretaceous volcanic arc, respectively, are characterized. In tones of brownish greenish, other sedimentary rocks from Neoauthocton cover are identified.

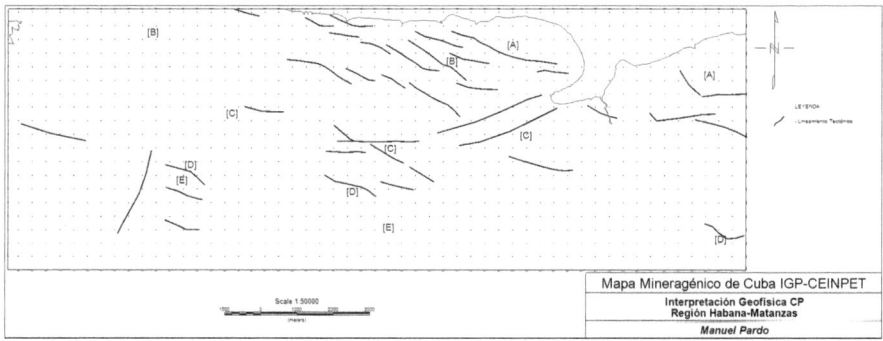

Fig. 4.13 Map of geophysical interpretation, Potential Fields, Habana-Matanzas region

At the regional level, the tectonic-structural decipherment of the study region can identify different elements from north to south (Fig. 4.13):

- The Northern Cuban Thrusts Belt, in the northeast end of the area, [A].
- The ophiolite belt of the northern region, [B].
- The central basin area (Aguacate-Mercedes), [C].
- The ophiolite belt of the southern region (Bejucal-Madruga elevations), [D].
- The southern basin area (Vegas), [E].

From the lithological mapping point of view, from the airborne gamma spectrometry data (Fig. 4.14), can be recognized in the ternary map of the three radioelements, weathering crusts of ophiolites, and associated volcanics (including related Chirino Fm.).

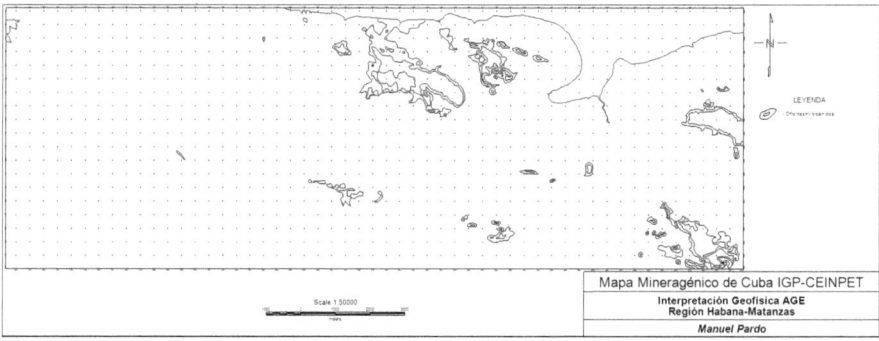

Fig. 4.14 Map of geophysical interpretation, Airborne Gamma Spectrometry, Habana-Matanzas region

4.4 Conclusions

As a result of the investigation, it offered a version of the tectonic-structural regionalization of territories under study based on potential fields, having as main references the map of the vertical derivative of the magnetic and gravimetric fields and together the analysis with the remaining transformations of these fields. In addition, it offers a version of the airborne gamma spectrometry mapping of weathering crusts corresponding to different magmatic and metamorphic lithological units. Thus, the chapter provides a summary of the main results achieved in the geological-structural mapping from potential fields and airborne gamma spectrometry data, of igneous and metamorphic rocks units in the western (Habana-Matanzas), central (Cienfuegos-Villa Clara-Sancti Spiritus) and central-eastern (Camagüey-Las Tunas-Holguín) regions of Cuba.

References

Colectivo de Autores (2010) Mapa Geológico Digital de la República de Cuba a escala 1:100,000. Inédito. Instituto de Geología y Paleontología, Servicio Geológico de Cuba, La Habana

Mondelo F, Sánchez R et al (2011) Mapas geofísicos regionales de gravimetría, magnetometría, intensidad y espectrometría gamma de la República de Cuba, escalas 1: 2,000,000 hasta 1:50,000. Inédito. IGP, La Habana, 278 p

Verduzco B, Fairhead JD, Green CM, McKenzie C (2004) New insights into magnetic derivatives for structural mapping. Lead Edge 23(2):116–119